Singapore Maths 2.0

The Stack Model Method

An Intuitive and Creative Approach to Solving Word Problems

Primary 3–4

S	m
T	e
A	t
C	h
K	o
	d

MATHPLUS Publishing

Yan Kow Cheong

MATHPLUS Publishing
Blk 639 Woodlands Ring Road
#02-35 Singapore 730639

E-mail: publisher@mathpluspublishing.com
Website: www.mathpluspublishing.com

First published in Singapore in 2016

National Library Board, Singapore Cataloguing-in-Publication Data

Yan, Kow Cheong, author.

The stack model method, Primary 3-4 : an intuitive and creative approach to solving word problems / Yan Kow Cheong. – Singapore : MathPlus Publishing, [2015]

pages cm. – (Stack model method ; 1)

ISBN : 978-981-09-4287-8

1. Problem solving – Study and teaching (Primary). 2. Mathematics – Study and teaching (Primary). 3. Mathematics – Problems, exercises, etc. I. Title. II. Series: Stack model method ; 1.

QA135.5
372.7044 -- dc23

OCN919071290

Printed in Singapore

Preface

You are familiar with the *Singapore Model Method*, or *Bar Method*, as a powerful visualisation problem-solving heuristic (or "problem-solving strategy", as maths teachers call it in the U.S.). What about the more intuitive and insightful Singapore's *Stack Model Method* for solving mathematics word problems?

No longer do you need to seek high and low to learn this insightful heuristic of modelling. Save your time and money; and most importantly, gain a competitive edge over your peers in mastering this visualisation heuristic in mathematical problem solving in just a few hours!

Learn what creative and innovative local maths teachers and tutors are teaching their above-average pupils. Update and upgrade yourself with commonly used problem-solving heuristics, such as the Stack Model Method to solve non-routine and challenging questions.

Disappointingly, because of maths editors' ignorance of the Stack Model Method, and poor marketing in promoting it, Singapore-published textbooks and workbooks, not to say, supplementary maths titles, have yet to be incorporated with this intuitive problem-solving heuristic. Moreover, you are unlikely to learn the Stack Model Method from attending Singapore maths workshops and conferences any time soon, or even from problem-solving courses offered to trainee or in-service teachers.

Stack up your visualisation skills by empowering yourself with the Singapore's stack model method! Like the Singapore's bar model method, the stack model method will help you to enhance your visual literacy or visualisation skills. The time to arm yourself mathematically and professionally is NOW!

As a Singapore maths educator, coach or parent, don't shortchange yourself by learning only a fraction of the Singapore maths curriculum; learn other problem-solving heuristics other than the Singapore model (or bar) method, which have until now not been made readily available in print and online. Here's your chance to learn the stack model method, a more powerful visualisation heuristic than the bar model method, which could help you and your pupils enhance both your creative and lateral thinking skills in primary mathematics.

Happy creative problem solving!

Yan Kow Cheong

Contents

A (*Very Short*) History of the *Stack Model Method*

Unlike the Singapore's "Model (or Bar) Method" that formally appeared in the mid-eighties, the "Stack Model Method", based on one or two self-published books, seems to have made its public appearance in Singapore about a decade ago.

Even today, a large proportion of local school maths teachers are not familiar with this powerful visualisation problem-solving heuristic, as compared to their counterparts teaching in enrichment and tuition centres, who started introducing it to primary 5–6 pupils, as an alternative method of solution to the bar model method.

Why the Stack Model Method has failed to reach out to a wider audience earlier is two-fold: (a) editors' unfamiliarity with the method; and (b) poor marketing efforts to promote the problem-solving heuristic in local schools. In other words, local publishers are clueless about publishing books on it, because their editorial staff were mostly non-maths majors, with a superficial understanding of even the bar method. In fact, about five years ago, I started submitting some solutions to word problems, which lend themselves to the Stack Model Method, but editors then conveniently rejected them — they showed zero interest to wanting to know how this visualisation heuristic works, or just refused to be mathematically or visually challenged.

Because the Stack Model Method was initially exemplified to solve primary 5–6 questions, primary maths teachers mistakenly thought that this problem-solving heuristic is only relevant to those teaching at these levels. In fact, as this book shows, the Stack Model Method can also be applied to solve non-routine questions in primary 3–4, or even at lower levels, especially when it comes to solving challenging or olympiad maths questions in primary 1–2.

In recent years, thanks to social media, it has become easier for local bloggers, writers and teachers to share with fellow maths educators online why the Stack Model Method is no longer an optional visualisation problem-solving heuristic, reserved only for the better or above-average maths pupils. There have been several requests to come up with a quick-and-dirty book on the Stack Model Method, but it was hard to publish it inexpensively, especially when local "maths editors" (and designers and typesetters) aren't prepared to have an intuitive feel of it.

After many excuses to put off writing a book on the Stack Model Method, and following more inquiries in recent years from maths educators overseas, who are keen to learn more about this intuitive and creative problem-solving method, I finally decided to write two quick-and-dirty books on this visualisation heuristic. The first book focusses on primary 3–4 questions; the second book looks at primary 4–6 questions, which are stack-model-friendly.

In the last revised editions of my local primary 3–6 assessment books, I was encouraged to witness that my maths editors this time round are more open and willing to incorporating, when space permits, a few alternative solutions that lend themselves to the stack model method. I am confident that the stack model method will be as popular as the bar model method in coming years, as both local and foreign maths educators are convinced why the stack model method is a more intuitive and creative problem-solving heuristic than the bar model method, especially when it comes to solving word problems that are bar-model-unfriendly.

Heuristic: Draw a diagram

Stack Method

(Singapore Maths 2.0)

Bar Method

(Singapore Maths 1.0)

Sakamoto Method

(Singapore Maths 2.0)

No Venn-ture, No gain

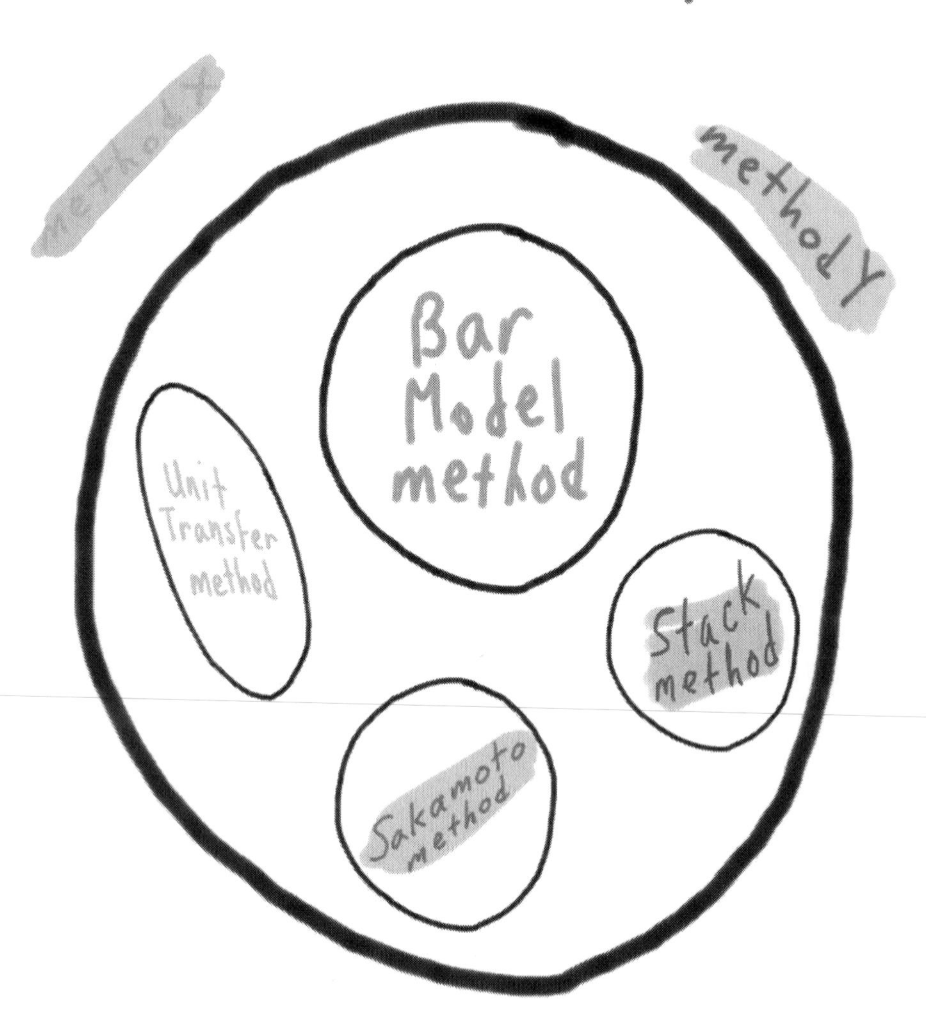

Bar method

Stack method

Baked in Singapore, not made in Singapore

Made in S'pore

Singapore MATHS

Singaporean MATHS

Think horizontal

THINK vertical

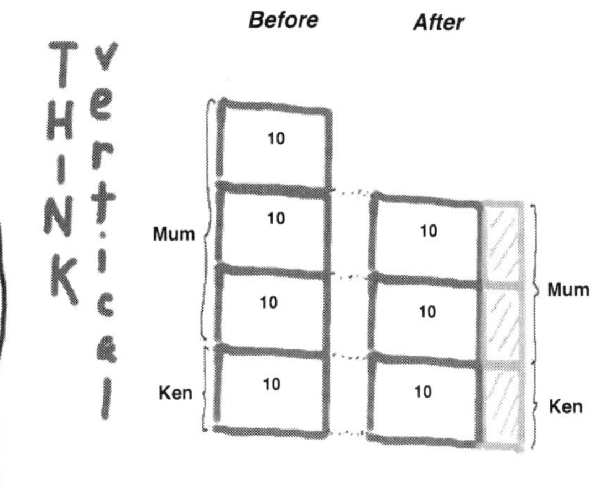

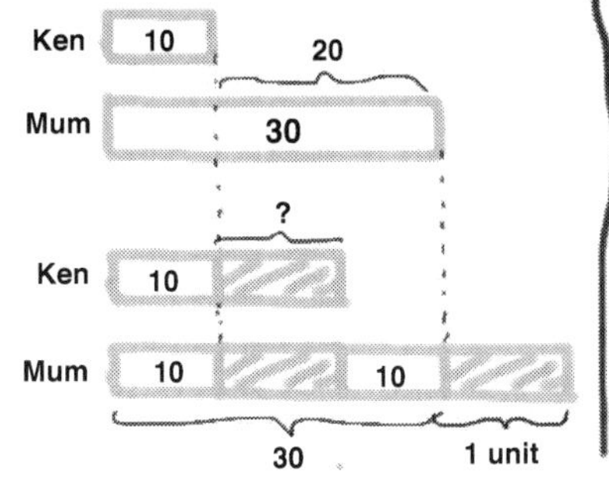

Bar method	Stack method
Hit SG ≈ 30 years	< 10 years old
Out of age	5 years late
Look-see proofs for ADULTS	Look-see proofs for KIDS
Reading a textbook (rows)	Reading a magazine (columns)

Are we not born-again STACKERS?

Simplistically speaking, the stack model method may be likened to adding a string of numbers vertically, as compared to the bar model method, which involves adding the same group of numbers horizontally.

12.34 + 567.8 + 9.01 = ?

12.34
567.8
+ 9.01
?

Which method is easier?

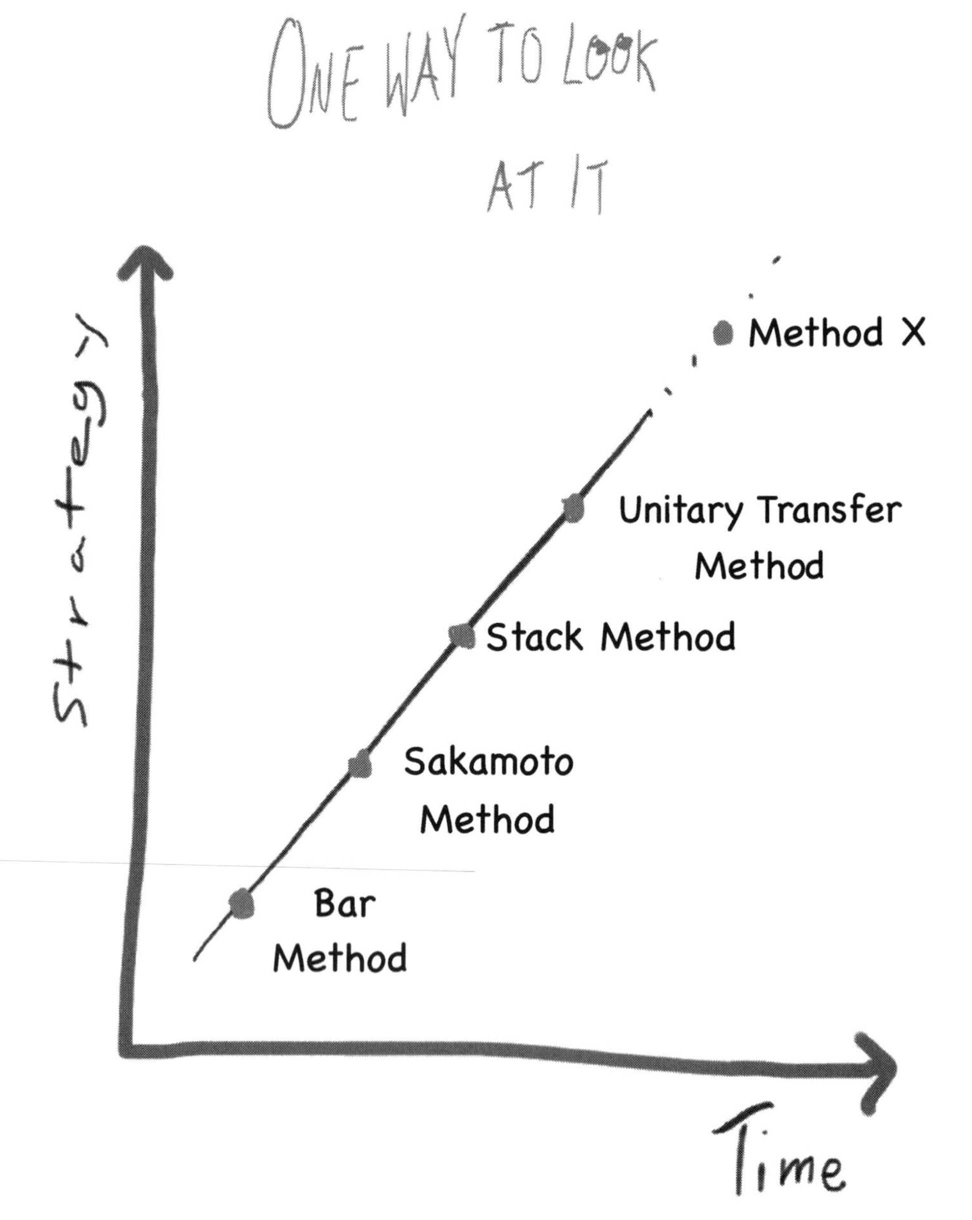
ONE WAY TO LOOK AT IT
Strategy
Method X
Unitary Transfer Method
Stack Method
Sakamoto Method
Bar Method
Time

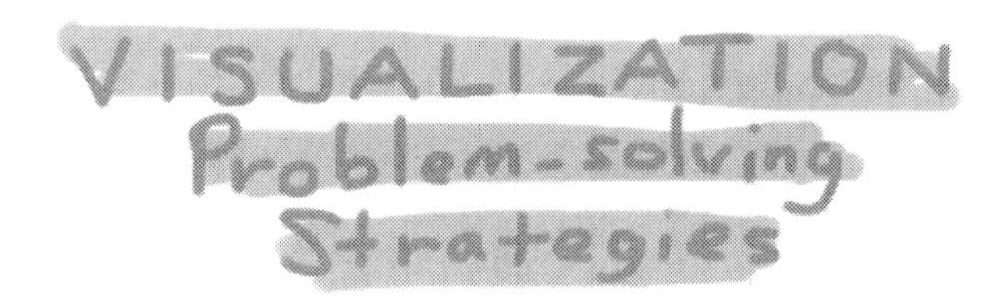

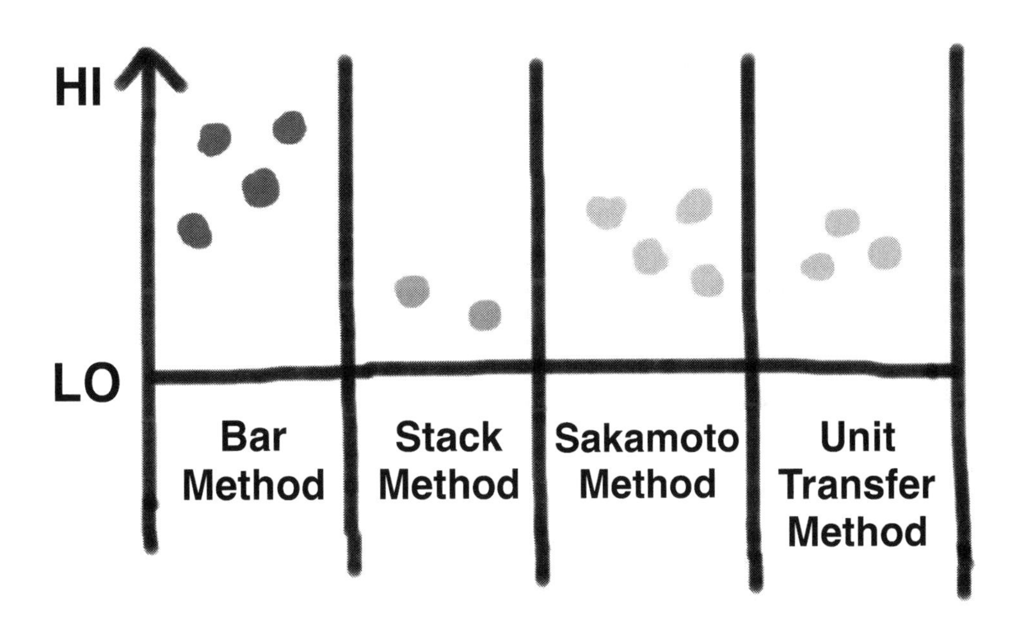

Singapore: Problem-Solving Heuristics
U.S.: Problem-Solving Strategies

What the Stack Method IS NOT

A STACK model **is not** an inverted BAR Model

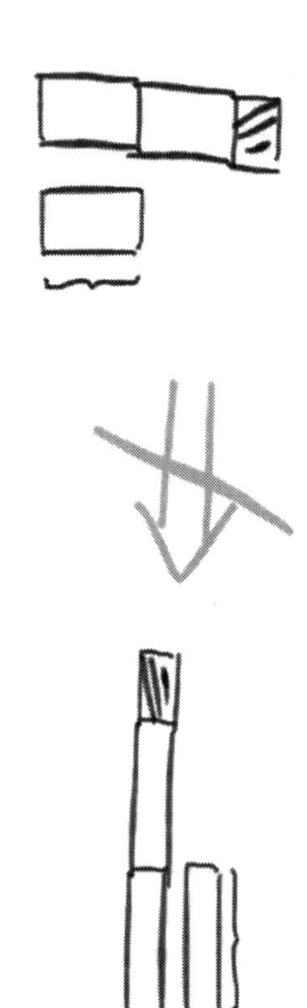

A devil's definition

Stack Model Method [maths]: A problem-solving heuristic that has been held hostage by pseudo-maths editors in the last decade, with Singapore maths authors and bloggers pushing for its release.

URL: www.stackmath.com

TWITTER: #stackmath

Futuring Robert Frost

Two model methods lay before me,
and I took the one less modelled.
And that has made all the difference.

Let's look at a "before-after" primary 4 question, taken from a Singapore maths assessment book.

Rick had 3 times as much money as Fiona. After Rick gave $285 to Fiona, he had twice as much money as she did. How much money did Rick have at first?

First, solve the above word problem, using the bar model method, then compare your solution with the stack-model solution below. *Can you figure out how the method works?*

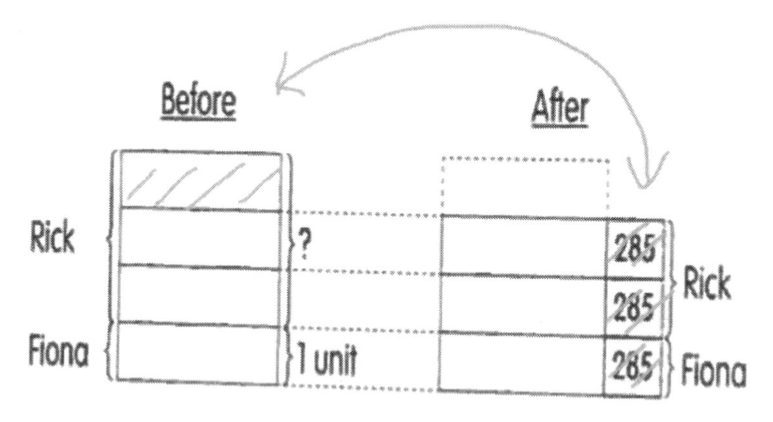

From the model,

1 unit ⟶ 3×285

3 units ⟶ $3 \times 3 \times 285 = 2565$

Rick had **$2565** at first.

Note: The total amount of money remains unchanged.

Singapore Maths 2.0

From Bar Method to Stack Method

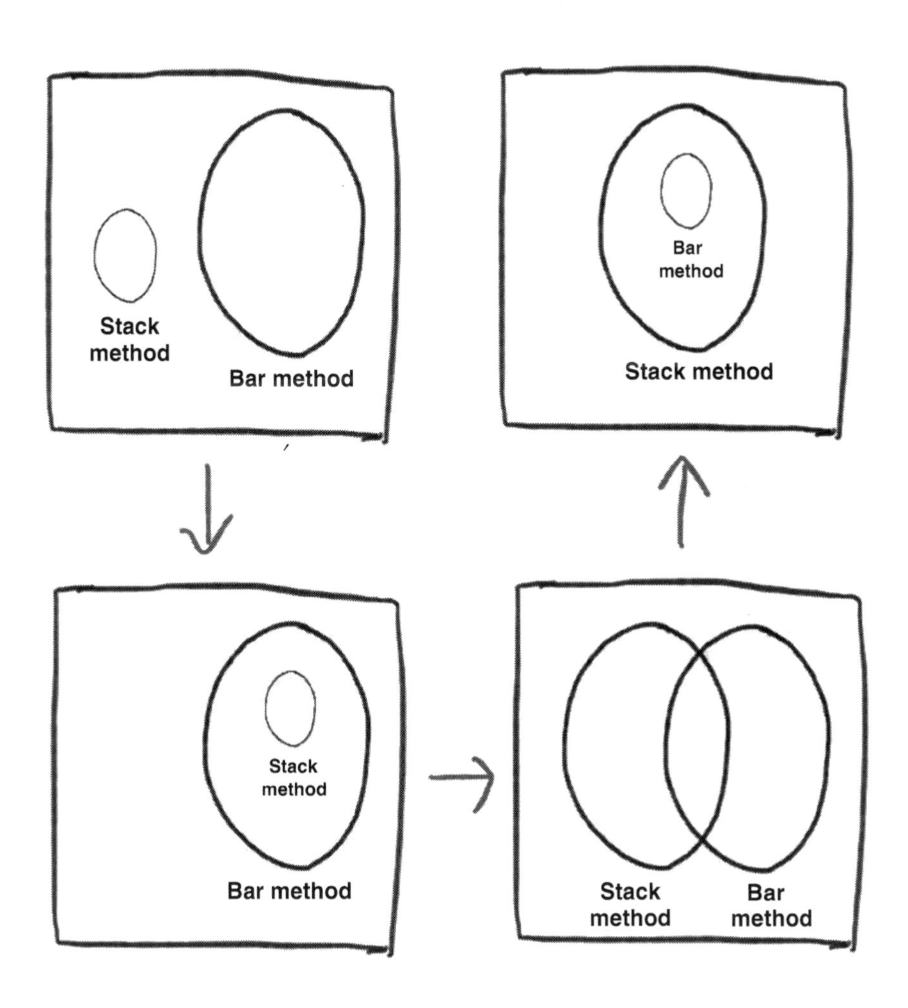

Basic Tools for

The Stack Method

Tools Needed:

Lines, Rectangles or Squares (and a flexible brain)

Stack models may be drawn horizontally or vertically — not diagonally or circularly yet!

4 units

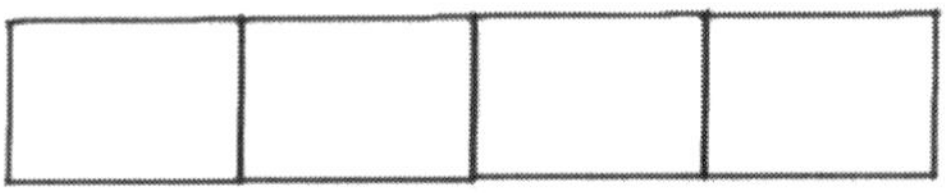

1 unit

4 units + 4

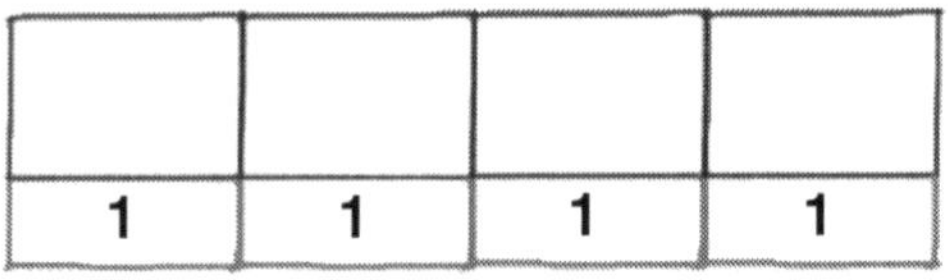

1 1 1 1

4 units – 4

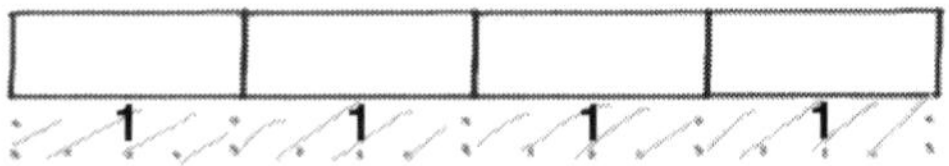

1 1 1 1

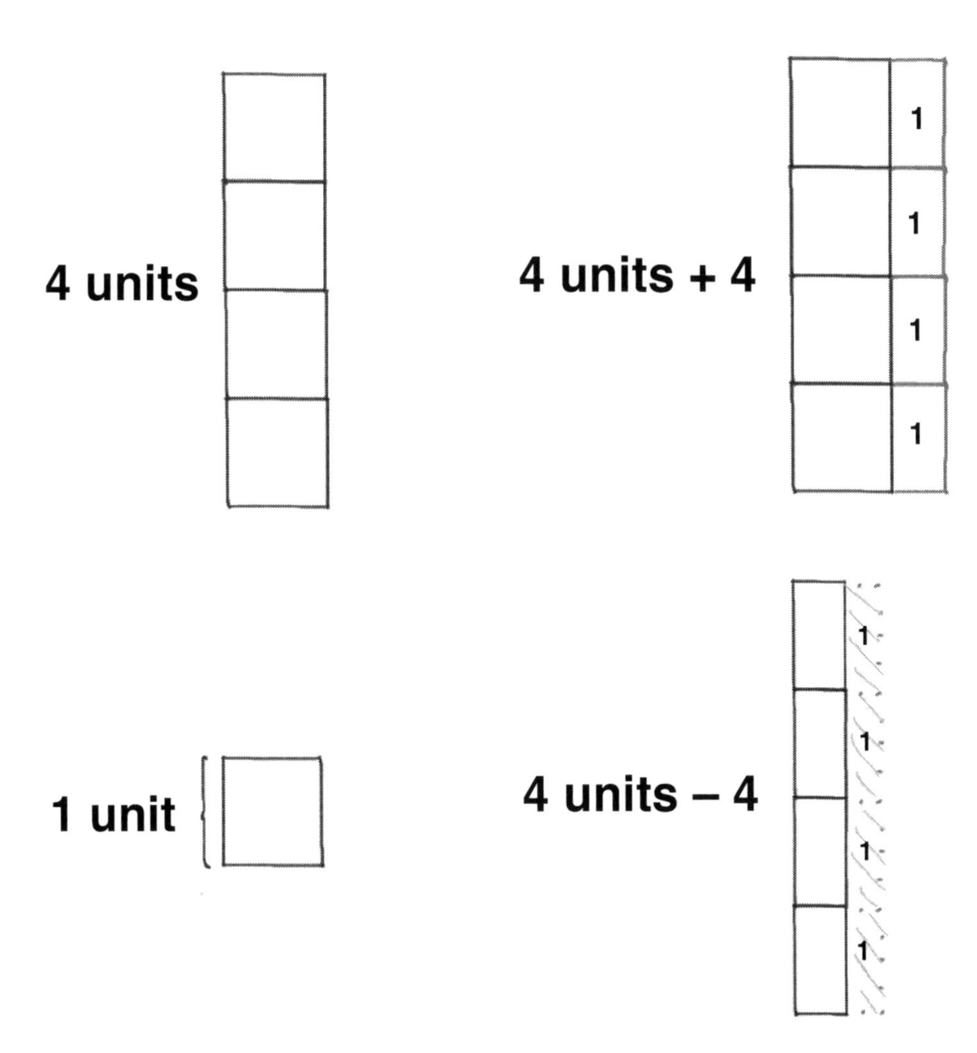

At first glance, it looks as if the stack model method is just the bar model method being rotated ninety degrees, but nothing is further from the truth. Although the bar model method and the stack model method share a fair bit in common, however, the thought process is different in each case.

Number Patterns

Number Patterns — Worked Examples 1–3

In primary 3–4, finding the sum of a series of numbers, such as the sum of a group of consecutive odd or even whole numbers, is usually performed, by looking out for pairs of numbers with the same sum.

This pairing technique may look like some “magic trick” for novice learners, without them understanding why this heuristic works. A “look-see” proof often helps to clear away any doubt, once pupils are able to see that “stacking up the numbers” in some particular way does the trick.

Worked Example 1

What is 1 + 2 + 3 + ... + 99 + 100?

Consider a simple case such as 1 + 2 + 3 + 4.

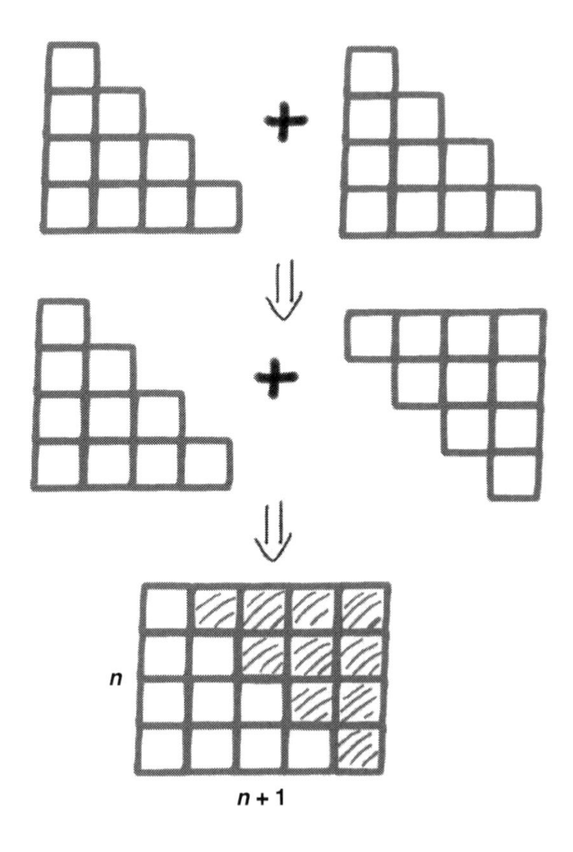

From the model, 1 + 2 + 3 + 4 = 1/2 × (4 × 5) = 1/2 × 20 = 10

For n = 100, 1 + 2 + 3 + ... + 99 + 100 = 1/2 × (100 × 101)
= 1/2 × 10 100
= 5050

Practice: What is 1 + 2 + 3 + ... + 59 + 60? Answer: 1830.

Alternatively, we may consider the numbers going up in a staircase from 1 to 100. The first step is one unit high, the second step is two steps high, the third step is three units high, up to the hundredth step, which is one hundred units high. So if we added up all the steps we would be adding up all the numbers from 1 to 100.

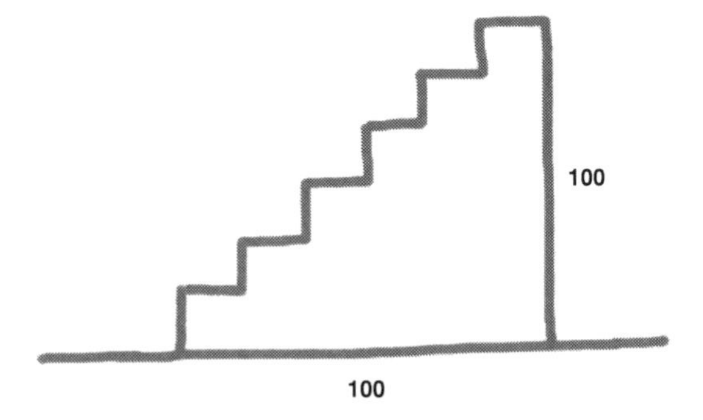

Now imagine a similar staircase placed upside-down over the first one. There has to be an overlap of one at the end in order to fit a similar staircase. What results is a rectangle which is 100 units along one side and 101 units along the other side. The total area is thus 100 × 101. That would give us "twice" the total we need because we added up "two" staircases so we divide by two. The answer is 5050.

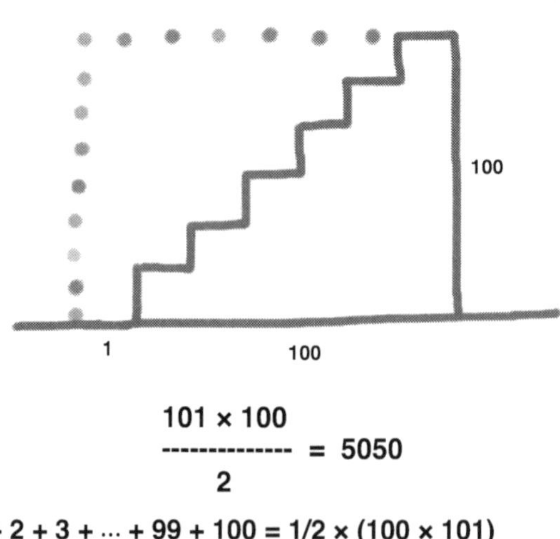

$$\frac{101 \times 100}{2} = 5050$$

$$\begin{aligned} 1 + 2 + 3 + \cdots + 99 + 100 &= 1/2 \times (100 \times 101) \\ &= 1/2 \times 10\,100 \\ &= 5050 \end{aligned}$$

Worked Example 2

What is the value of 1 + 3 + 5 + 7 + ... + 97 + 99?

Method 1

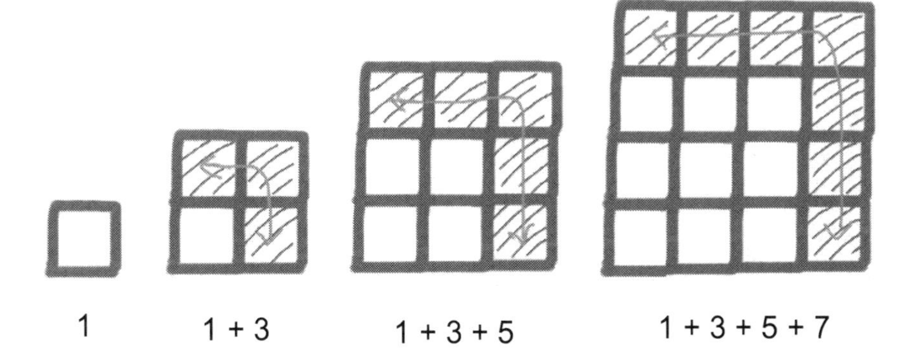

Clearly, adding together the first n odd terms gives a sum of n^2.

1 = 1
1 + 3 = 4 = 2 × 2
1 + 3 + 5 = 9 = 3 × 3
...
1 + 3 + 5 + ... + 97 + 99 = 50 × 50 = 2500

What is the value of 1 + 3 + 5 + 7 + ··· + 97 + 99?

Method 2

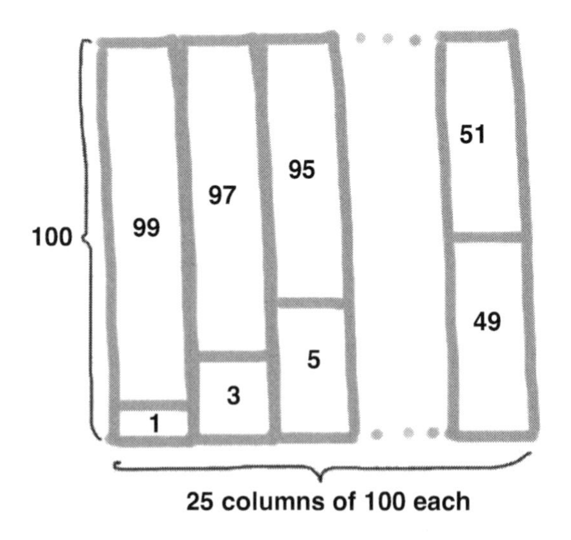

From the model,

1 + 3 + 5 + 7 + ... + 97 + 99 = 25 × 100 = 2500

Practice

What is 1 + 3 + 5 + ... + 39 + 41?

Answer: 441.

Worked Example 3

What is the value of 2 + 4 + 6 + ⋯ + 98 + 100?

Method 1

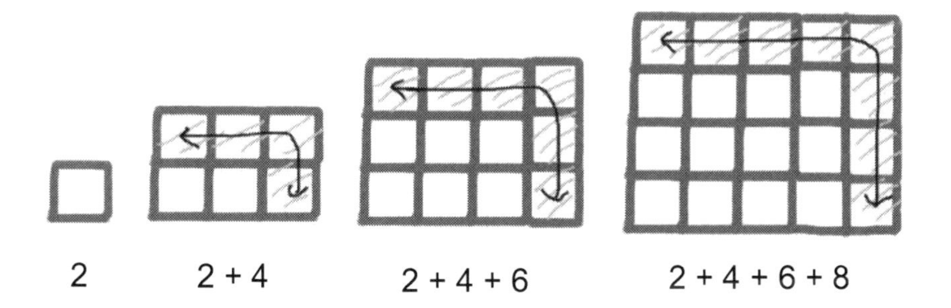

$2 = 1 \times 2$
$2 + 4 = 2 \times 3$
$2 + 4 + 6 = 3 \times 4$
…

In general, the nth sum will yield a rectangle with dimensions $n \times (n + 1)$.

$2 + 4 + 6 + \dots + 98 + 100 = 50 \times 51 = 2550$

What is the value of 2 + 4 + 6 + ... + 98 + 100?

Method 2

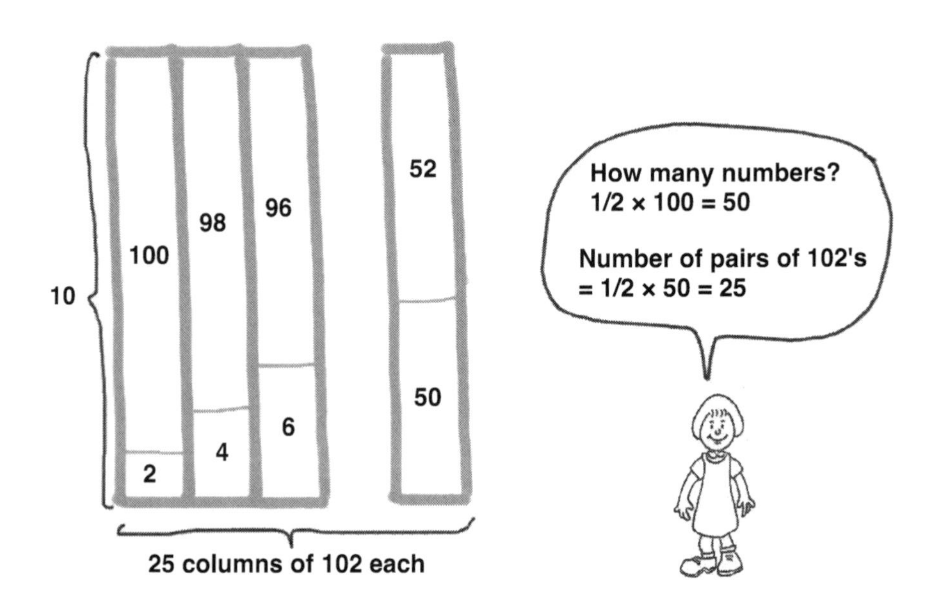

From the model,

$2 + 4 + 6 + \dots + 98 + 100 = 25 \times 102 = 2550$

Practice

What is 2 + 4 + 6 + ... + 68 + 70?

Answer: 1260.

Whole Numbers

Whole Numbers — Worked Examples 4–16

The examples here show that, unlike bar modelling, stack modelling often helps to solve a number of both routine and non-routine questions, by using only one model drawing, instead of constructing two or more models to make the comparison.

Often times, we can find what is required at one go, by short-circuiting some intermediate result that may be needed if we resort to bar modelling. In other words, stack modelling helps us to better "see" how the parts are related to the whole, because of the way the stack model is designed.

Stack modelling allows for more flexibility, as compared to bar modelling, which tends to be traditionally done in a horizontal manner, at least this is what most students and teachers seem to believe from a mere casual reading of modelled solutions from assessment maths books.

Worked Examples 12 and 13 serve as a "look-see" proof for the "Make a supposition" heuristic (or strategy, as local teachers in the U.S. call it). Primary 3–4 pupils are often puzzled why the "Make a supposition" heuristic works, although a number of them know how to apply it effectively, besides the guess-and-check heuristic.

Worked Example 4

**I am thinking of two numbers.
Their sum is 1544 and their difference is 152.
What are the two numbers?**

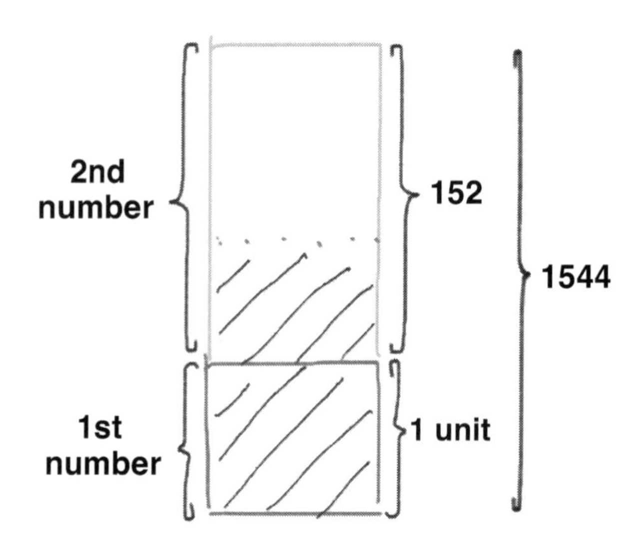

From the model,

2 units → 1544 – 152 = 1392

1 unit → 1392 ÷ 2 = 696

The smaller number is 696.

696 + 152 = 848 or 1544 – 695 = 848

The greater number is 848.

Thought Process

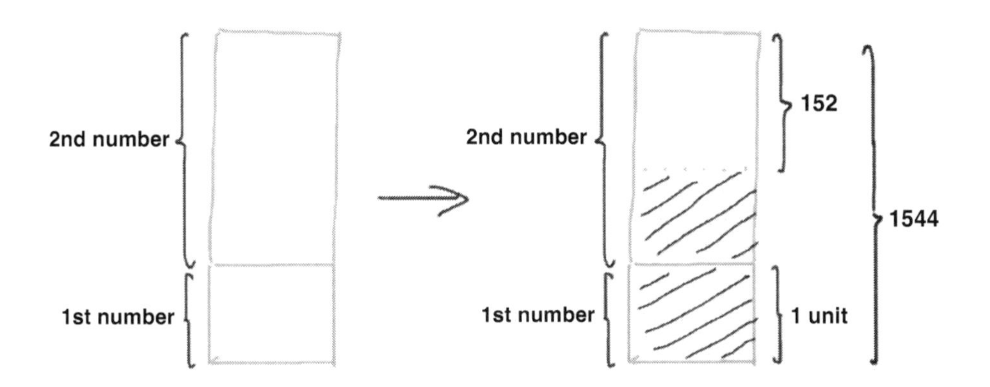

Alternatively, we may solve the problem as follows:

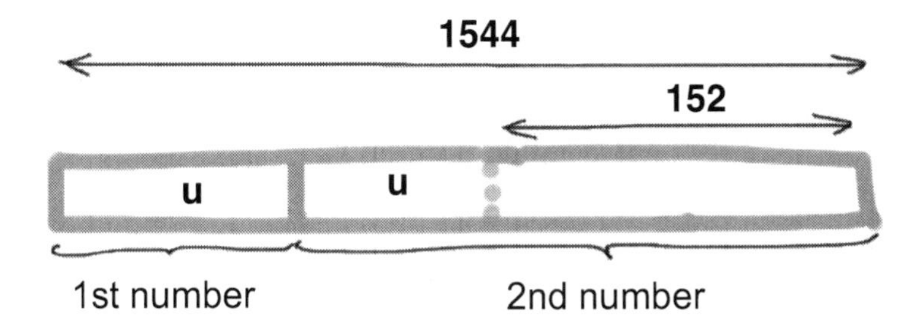

From the model,

2u → 1544 – 152 = 1392
u → 1392 ÷ 2 = 696

The smaller number is 696.

696 + 152 = 848 or 1544 – 695 = 848
The greater number is 848.

Note: In bar modelling, you would need to draw two bars of different lengths to compare the two numbers, then decide which one of the two bars would be defined as one unit, before forming a relationship between the units and the quantities involved.

In stack modelling, only one bar is needed to compare the two numbers, which is more efficient and faster.

Practice

Joel dreamed of two numbers whose sum is 1234 and whose difference is 258. Which two numbers did he dream of?

Answer: 488, 746.

Worked Example 5

Gregory has 48 stamps. Steven has 26 stamps. How many stamps must Gregory give to Steven so that they have the same number of stamps? How many stamps will each of them have now?

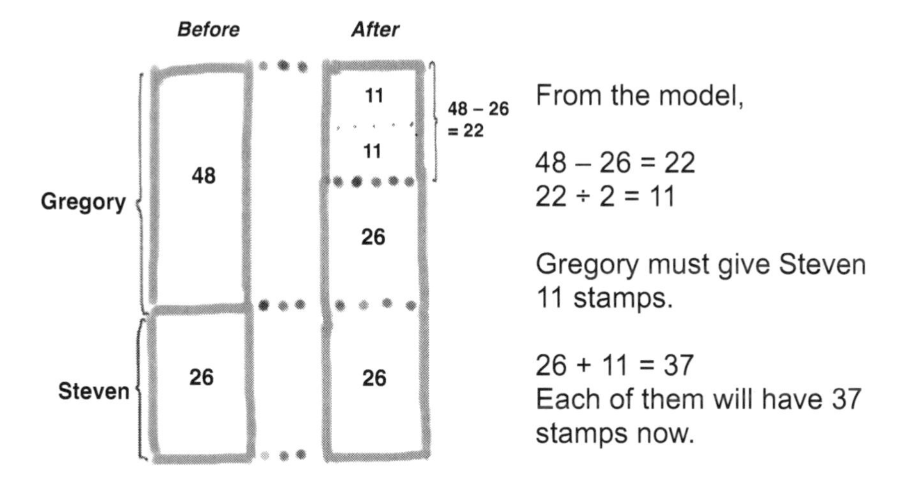

From the model,

48 – 26 = 22
22 ÷ 2 = 11

Gregory must give Steven 11 stamps.

26 + 11 = 37
Each of them will have 37 stamps now.

Alternatively, since the total number of stamps does not change, and we are only asked how many stamps each person will get in the end, we could also solve the question as follows:

48 + 26 = 74
74 ÷ 2 = 37
Each person will then have 37 stamps.

Practice

Doris has 56 postcards and Mark has 38 postcards. How many postcards must Doris give to Mark so that they have the same number of postcards? How many postcards will each of them have now? Answer: 9.

Worked Example 6

Ryan has 500 more stamps than Oliver at first. He gives 300 stamps to Oliver. Who has more stamps now? How many more?

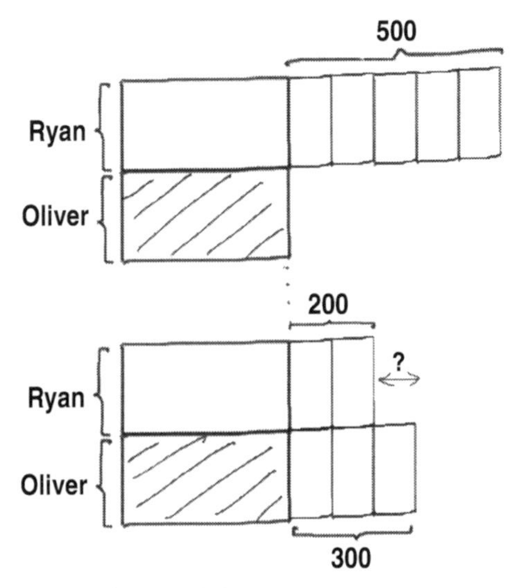

From the model,

500 – 300 = 200
300 – 200 = 100

Oliver has 100 more stamps than Ryan now.

A common mistake is to take 500 – 300 = 200, and to claim that Ryan has 200 more stamps than Oliver.

Practice

Laval has 400 more books than Idris. Laval gives 300 books to Isris. Who has more books now? How many more?

Answer: Idris; 200 books.

Worked Example 7

Paula has 350 fewer coins than Doris. Doris gives 250 coins to Paula. Who has fewer coins now? How many fewer?

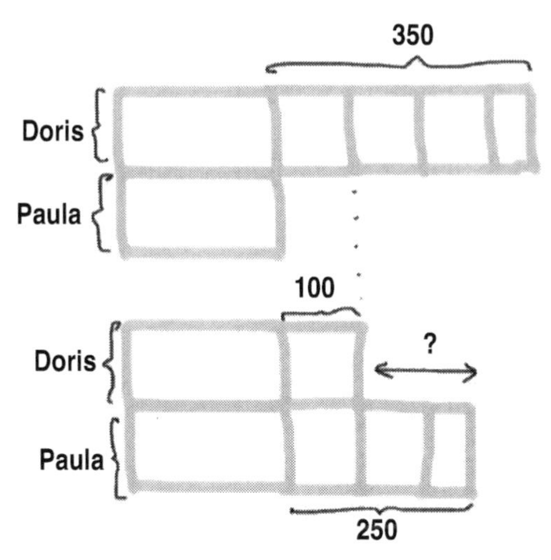

From the model,

350 – 250 = 100
250 – 100 = 150

A common mistake is to say that Paula had 350 – 250 = 100 fewer coins than Doris.

Doris has 150 fewer coins than Paula now.

Practice

Rajoo has 250 fewer stamps than Joan. Rajoo receives 150 stamps from Joan. Who has fewer stamps now? How many fewer?

Answer: Joan; 50 stamps.

Why *Stack Modelling*?

In many instances, a question may be solved by the bar or stack method. In fact, for easier or routine questions, there is no conceptual advantage to choosing one method over the other.

However, as the questions get more difficult, especially when the problems don't lend themselves easily to the bar method, we find that in a number of cases, the stack method offers a more intuitive method of solution than the bar method, particularly for questions from traditionally deemed-difficult primary school maths topics like *Fraction*, *Ratio* and *Percentage*.

Meanwhile, one natural way to get pupils to be familiar with the stack method as a powerful problem-solving heuristic in solving word problems, is to have them apply it to solve both routine and non-routine questions, as early as in primary 3 and 4. By the time they reach primary 5 and 6, the stack model method will not look like a foreign problem-solving heuristic anymore *vis-à-vis* the bar model method — the heuristic commonly used in current Singapore maths textbooks and workbooks.

Stack modelling is not a competitor of bar modelling; in fact, one who is already versed with the ins and outs of bar modelling, would better appreciate the power of stack modelling, as one could see why it is preferable to use the stack model over the bar model in a number of cases. Bar and stack models complement each other, with the stack method being at its best showing when the questions get complex and computationally tedious.

Worked Example 8

There are 2450 adults at an assembly. 896 of them are women. How many more men than women are there?

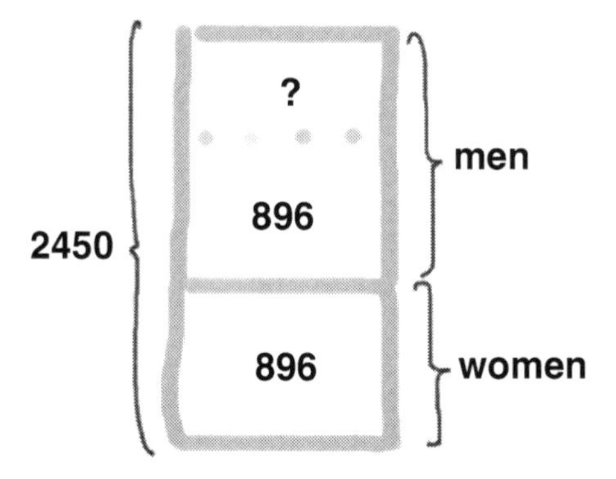

From the model,

Note: There is no need to calculate the number of men first to find the answer.

$$\begin{aligned}&2450 - 896 - 896\\ &= 2450 - 900 - 900 + 4 + 4\\ &= 2450 - 1800 + 8\\ &= 650 + 8\\ &= 658\end{aligned}$$

There are 658 more men than women.

Observe that in bar modelling, a common approach is to draw two bars, then determine the number of men, before finding the difference between men and women to find the answer.

Practice

Bob and Ian have $52.80 altogether. If Ian has $18.90, how much more money does Bob have than Ian?

Answer: $15.

Worked Example 9

A baker sold a total of 1320 loaves of bread in June and July. He sold 678 loaves in June and 901 loaves in August. How many more loaves did he sell in August than in July?

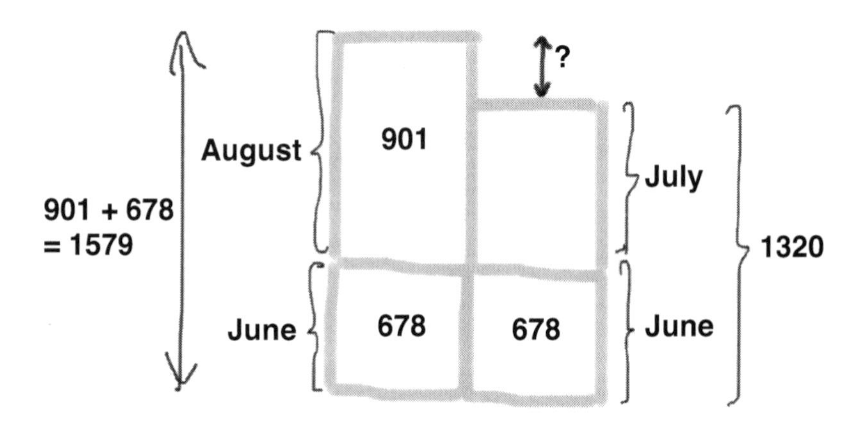

From the model,

1579 – 1320 = 259

He sold 259 more loaves in August than in July.

Alternatively, we can find the number of loaves sold in July first, before taking the difference between the number of loaves sold in July and August.

Worked Example 10

Sally had 57 more pencils than pens. After she donated 47 pencils, she had twice as many pencils as pens. How many pens and pencils did she have left altogether?

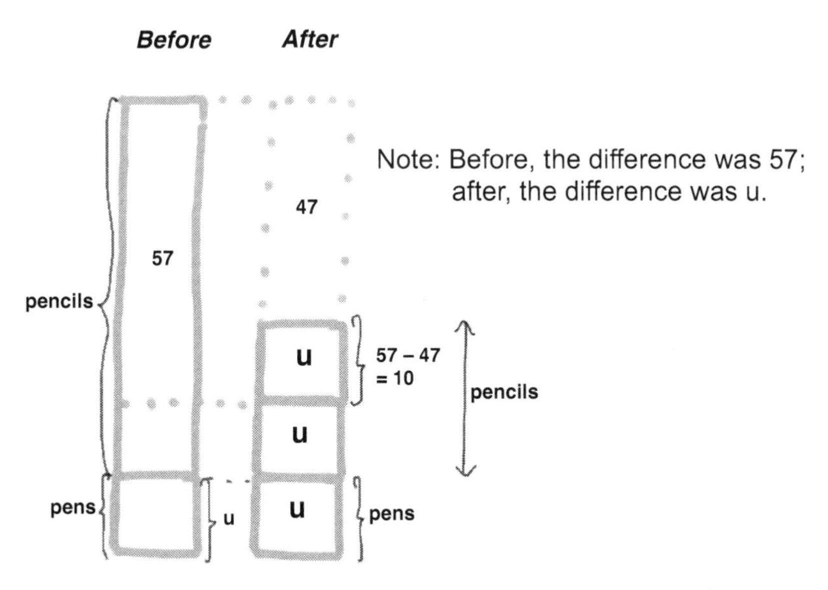

From the model,

u → 57 – 47 = 10
3u → 3 × 10 = 30

Sally had 30 pens and pencils left altogether.

Practice

Eunice had 48 more Singapore coins than Malaysia coins. After giving away 16 Singapore coins, she had twice as many Singapore coins as Malaysia coins. How many Singapore and Malaysia coins did she have left in all? Answer: 96 coins.

Worked Example 11

Gerald had 59 fewer stamps than coins. After she donated 23 coins, he had 3 times as many coins as stamps. How many stamps and coins did he have left altogether?

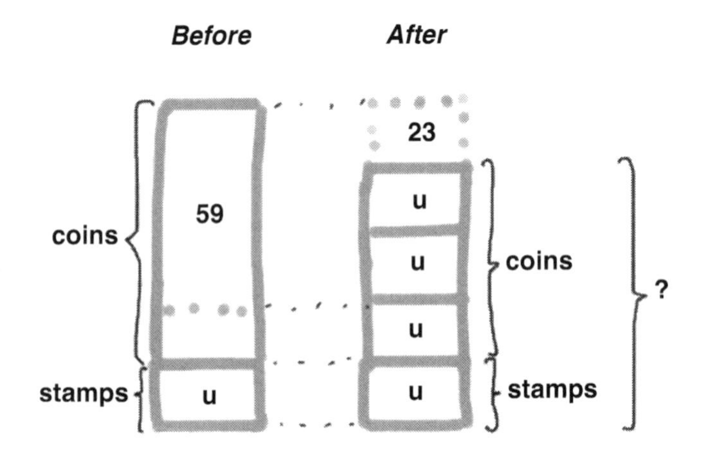

From the model,

Note: Before, the difference was 59; after, the difference was 2u.

2u → 59 – 23 = 36
4u → 36 + 36 = 72

Gerald had 72 stamps and coins left.

Practice

Lisa had 44 fewer red buttons than blue buttons. After using up 17 blue buttons, she had 4 times as many blue buttons as red buttons. How many red and blue buttons did she have left in total? Answer: 45 buttons.

Albert saw 15 wild cats and flamingoes at the zoo. A wild cat has 4 legs and a flamingo has 2 legs. He counted their legs and found out that they have 44 legs altogether. How many wild cats and how many flamingoes did he see at the zoo?

Two commonly used non-visual heuristics by teachers and pupils are:

Method 1 (Guess and check)

A wild cat has 4 legs.
A flamingo has 2 legs.

Number of wild cats	Number of flamingoes	Total number of legs
9	6	$9 \times 4 + 6 \times 2$ $= 36 + 12$ $= 48$ (Too high)
8	7	$8 \times 4 + 7 \times 2$ $= 32 + 14$ $= 46$ (Too high)
(7)	8	$7 \times 4 + 8 \times 2$ $= 28 + 16$ $= 44$ (Correct)

He saw **7** wild cats at the zoo.

Method 2 (Make a supposition)

Suppose there were a total of 15 flamingoes
Then the total number of legs would be
$15 \times 2 = 30$.

But there are a total of 44 legs, so the extra $44 - 30 = 14$ legs must have come from the wild cats.

One wild cat has 2 more legs than one flamingo.

Thus there must be $14 \div 2 = 7$ wild cats.

He saw **7** wild cats at the zoo.

Visualising the "Make a Supposition" Heuristic

Worked Example 12

Albert saw 15 wild cats and flamingoes at the zoo. A wild cat has 4 legs and a flamingo has 2 legs. He counted their legs and found out that they have 44 legs altogether. How many wild cats and how many flamingoes did he see at the zoo?

Method 1

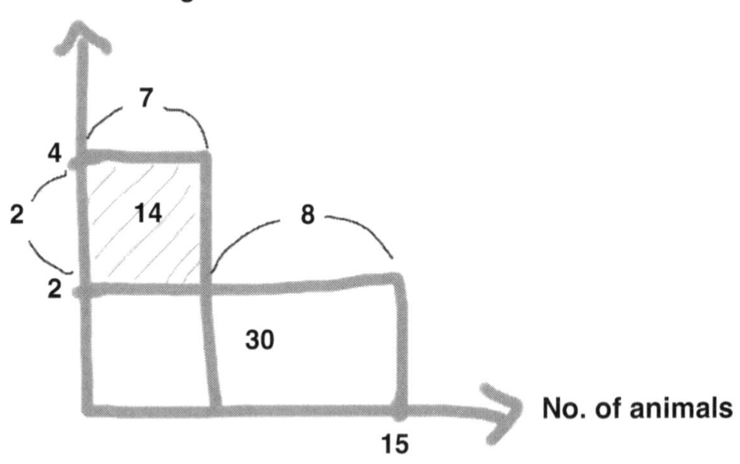

15 × 2 = 30 (if all 15 animals were flamingoes)

44 – 30 = 14
4 – 2 = 2 (difference in the number of legs)

14 ÷ 2 = 7 → wild cats
15 – 7 = 8 → flamingoes

Albert saw 7 wild cats and 8 flamingoes at the zoo.

Thought Process

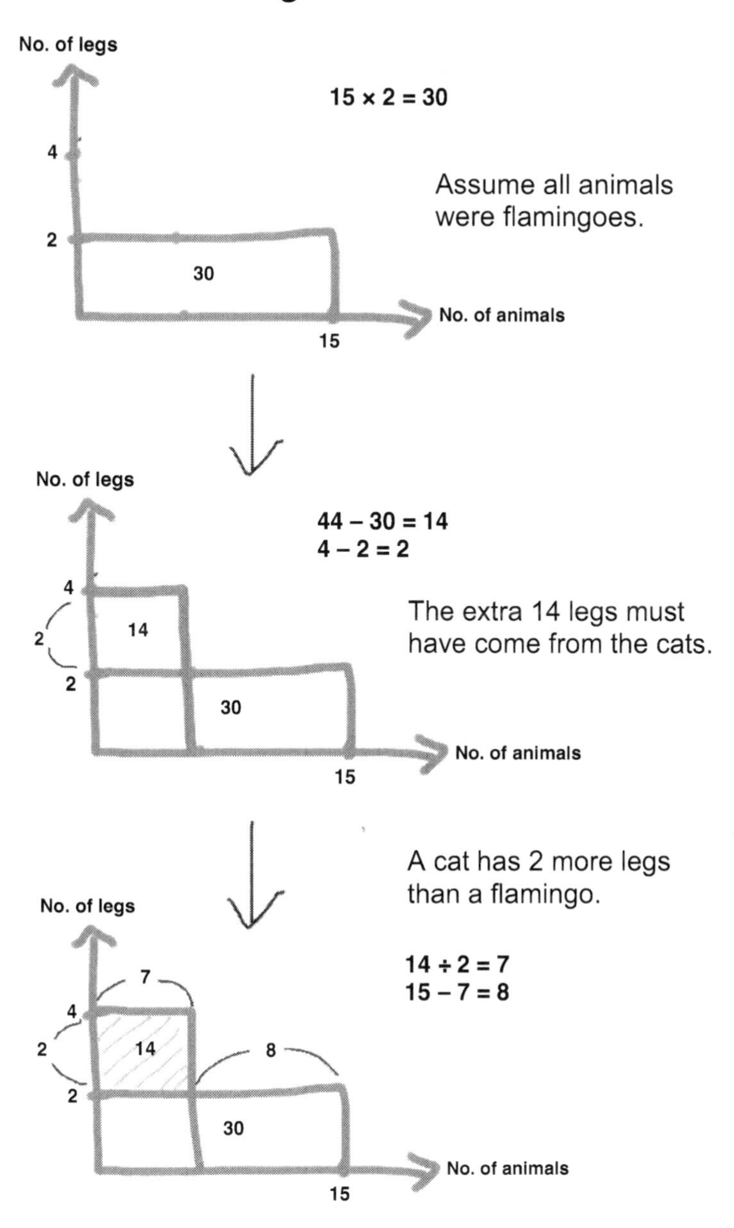

Albert saw 15 wild cats and flamingoes at the zoo. A wild cat has 4 legs and a flamingo has 2 legs. He counted their legs and found out that they have 44 legs altogether. How many wild cats and how many flamingoes did he see at the zoo?

Method 2

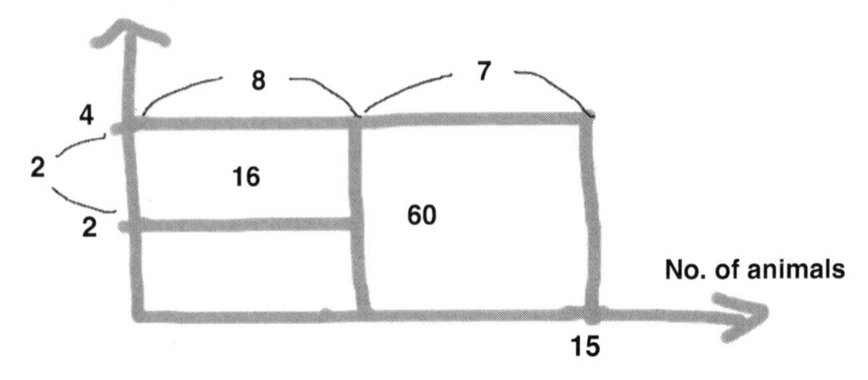

$15 \times 4 = 60$ (if all animals were wild cats)

$60 - 44 = 16$

$4 - 2 = 2$ (difference in the number of legs)
$16 \div 2 = 8 \rightarrow$ flamingoes
$15 - 8 = 7 \rightarrow$ wild cats

Albert saw 7 wild cats and 8 flamingoes at the zoo.

Thought Process

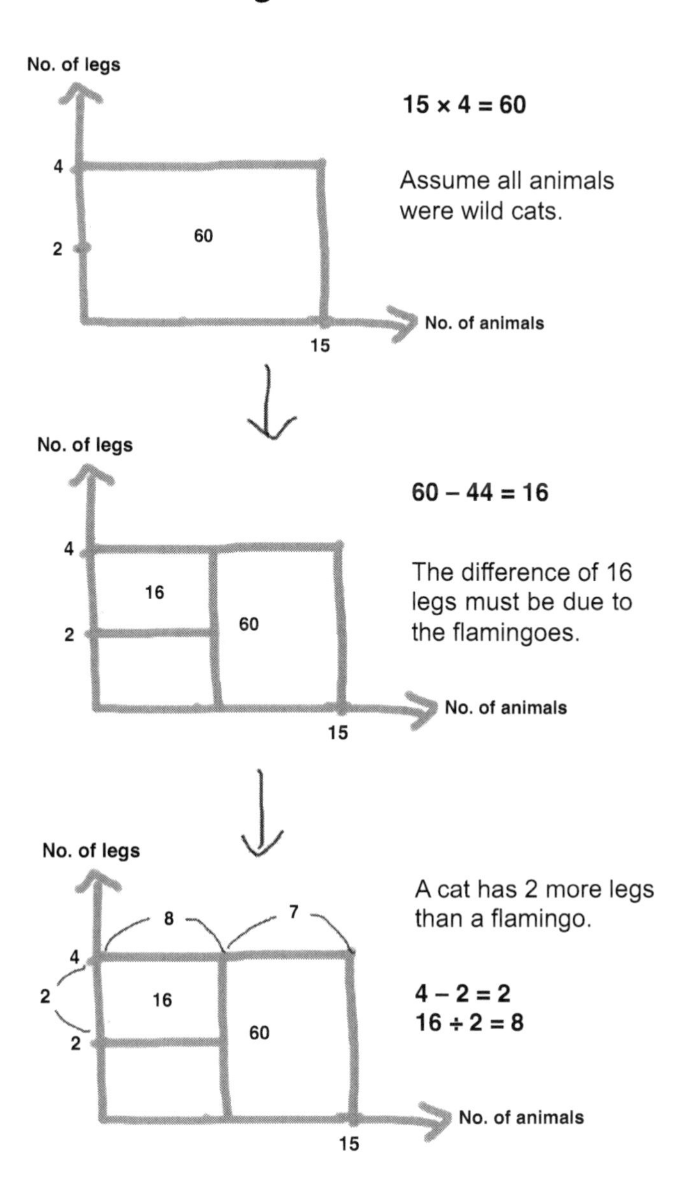

15 × 4 = 60

Assume all animals were wild cats.

60 – 44 = 16

The difference of 16 legs must be due to the flamingoes.

A cat has 2 more legs than a flamingo.

4 – 2 = 2
16 ÷ 2 = 8

Worked Example 13

Farmer Yan has 50 chickens and rabbits.
The total number of legs is 128.
How many chickens and how many rabbits are there?

Method 1

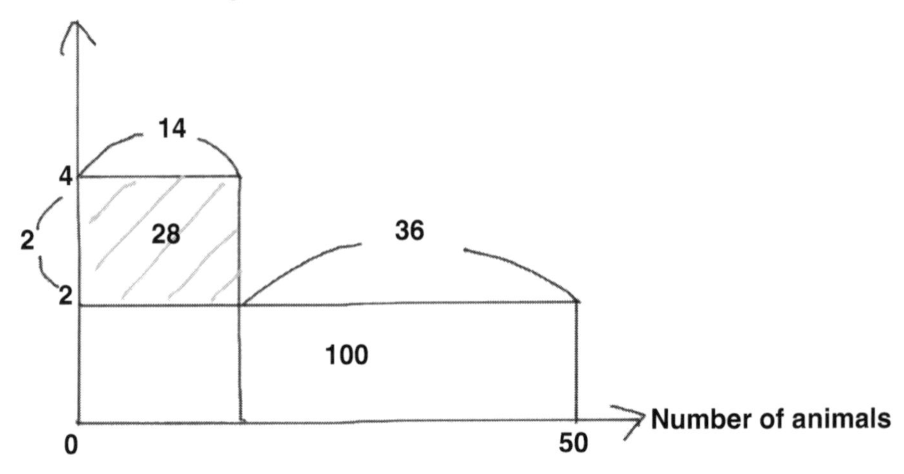

50 × 2 = 100 (if all animals were chickens)

128 – 100 = 28
4 – 2 = 2 (difference in the number of legs)

28 ÷ 2 = 14 → rabbits
50 – 14 = 36 → chickens

There are 36 chickens and 14 rabbits.

Method 2

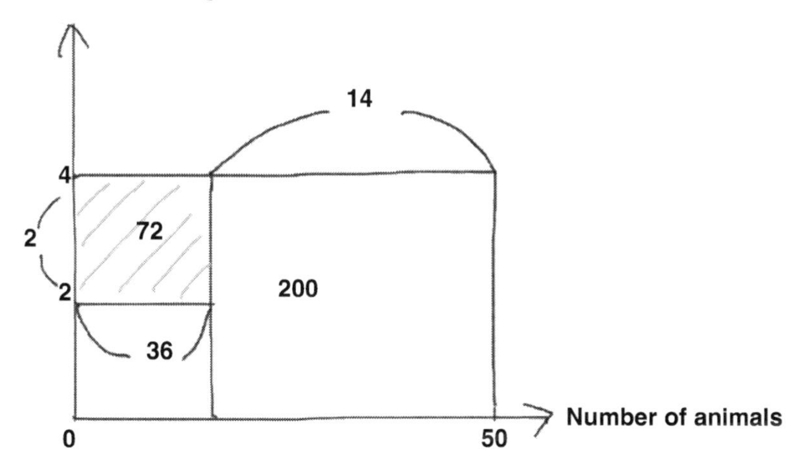

50 × 4 = 200 (if all animals were rabbits)

200 – 128 = 72
4 – 2 = 2 (difference in the number of legs)

72 ÷ 2 = 36 → chickens
50 – 36 = 14 → rabbits

There are 36 chickens and 14 rabbits.

Practice

A bike's owner has 30 bicycles and tricycles. The total number of wheels is 73. How many bicycles and how many tricycles does he have?

Answer: 17 bicycles, 13 tricycles.

Worked Example 14

Mrs Smith goes into a store and orders 85 blue balloons and green balloons for her daughter's birthday. She wants 25 more green balloons than blue balloons. How many balloons of each colour will she take home?

Method 1

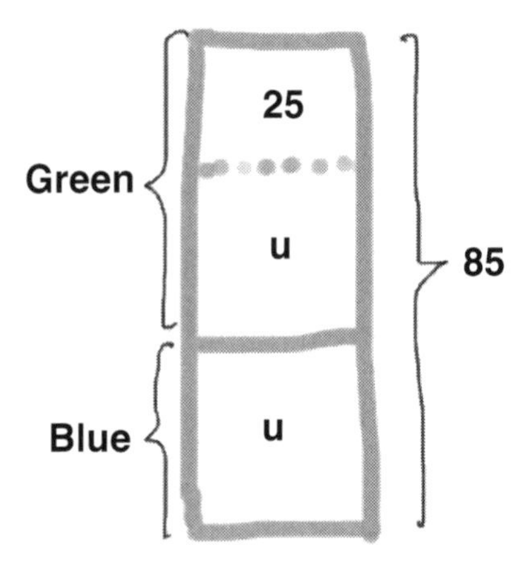

From the model,

2u → 85 – 25 = 60
u → 60 ÷ 2 = 30
u + 25 → 30 + 25 = 55 or 85 – 30 = 55

She will take home 30 blue and 55 green balloons.

Worked Example 14

Mrs Smith goes into a store and orders 85 blue balloons and green balloons for her daughter's birthday. She wants 25 more green balloons than blue balloons. How many balloons of each colour will she take home?

Method 2

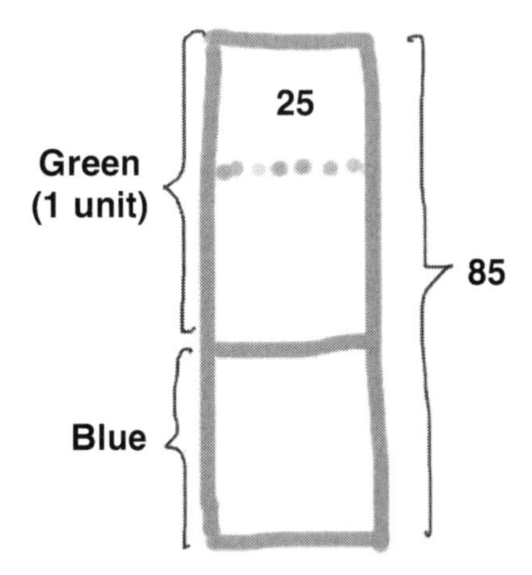

From the model,

2 units → 85 + 25 = 110
1 unit → 110 ÷ 2 = 55
1 unit – 25 → 55 – 25 = 30 or 85 – 55 = 30

She will take home 30 blue and 55 green balloons.

Worked Example 15

Roy and June have $137 altogether. June has $29 more than Roy. How much money does each of them have?

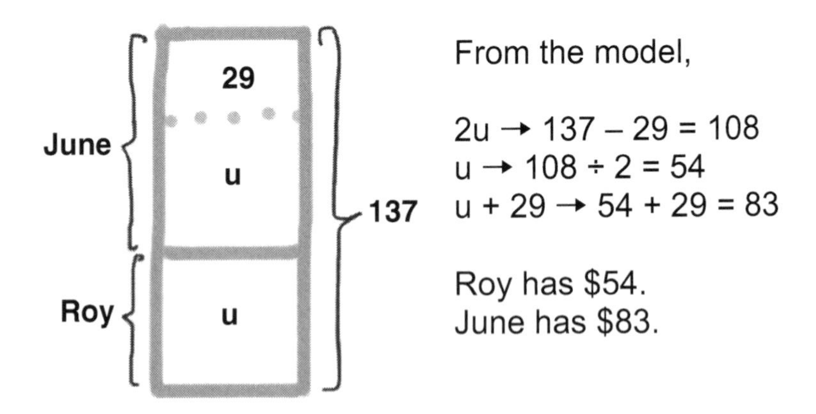

From the model,

$2u \rightarrow 137 - 29 = 108$
$u \rightarrow 108 \div 2 = 54$
$u + 29 \rightarrow 54 + 29 = 83$

Roy has $54.
June has $83.

Alternatively, we could also solve the question, by first assigning June's instead of Roy's amount of money to be "one unit".

Practice

In total, Aaron and Carl have $58.10. Aaron has $9.90 less than Carl. How much money does each of them have?

Hint: Convert dollars to cents.

Answer: Aaron: $24.10; Carl: $34.

Worked Example 16

The sum of the number of e-books Bob and Paul have is 30. If Paul deleted 5 of his e-books and Bob doubled the number of e-books he has, they would then have a total of 46 e-books. How many e-books does each have?

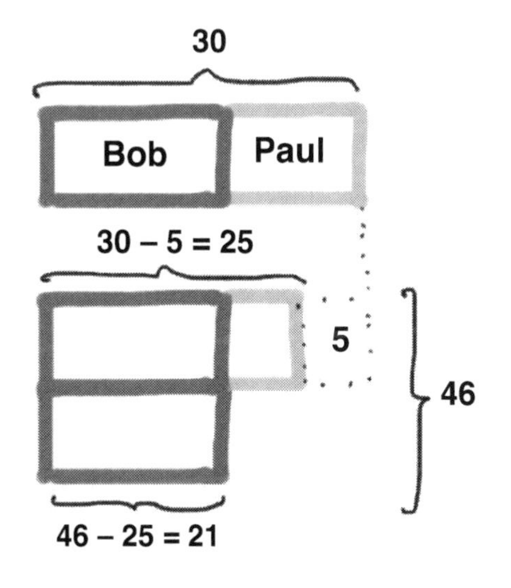

From the model,

46 – 25 = 21
Bob has 21 e-books.

30 – 21 = 9
Paul has 9 e-books.

Check:
Before
Bob: 21; Paul: 9
After
Bob: 42; Paul: 9 – 5 = 4
42 + 4 = 46

Alternatively, we may solve the question as follows:

2 × Bob + Paul → 46 + 5 = 51
Given: Bob + Paul → 30
Bob → 51 – 30 = 21
Paul → 30 – 21 = 9

Bob has 21 e-books.
Paul has 9 e-books.

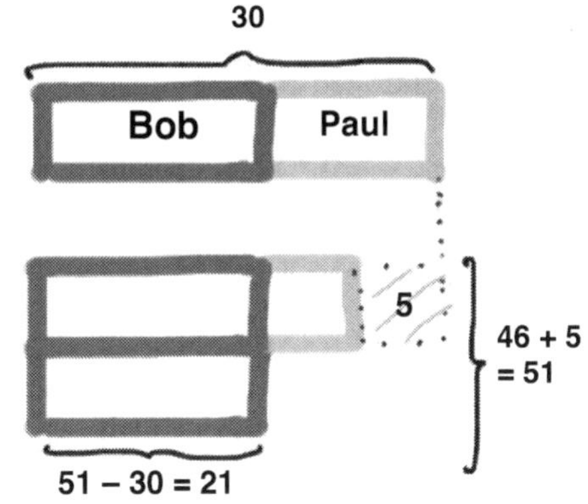

Age Problems

Age Problems — Worked Examples 17–20

Although age-related problems can be solved by bar modelling, however, a number of them can prove to be quite a challenge, even for above-average pupils and teachers. Constructing a user-friendly bar model that even an average maths pupil can see with his mind's eye is not that easy, unless one is fluent with model-drawing — *not all bar models are created equal.*

Surprisingly, a number of these brain-unfriendly age problems that are bar-model-unfriendly lend themselves quite easily to the stack model. This is because of the way the units or parts are stacked up, which makes it easier for one to see the numerical relationship between the quantities or unknowns involved, as compared to the same units or parts being arranged horizontally in a bar model.

Several stack-model solutions are offered for a given problem, as an incentive to encourage readers to think creatively and laterally. Or, given the same stack model, how can one apply different thought processes to arrive at the answer?

Stack modelling allows age-related problems that used to be discussed at higher levels to be set in lower levels, as the visualisation skills needed to solve these problems are within the ability of most primary 3–4 pupils.

Worked Example 17

Karen is 10 years old and her mother is 30 years old. How many years later will Karen's mother be twice as old as her?

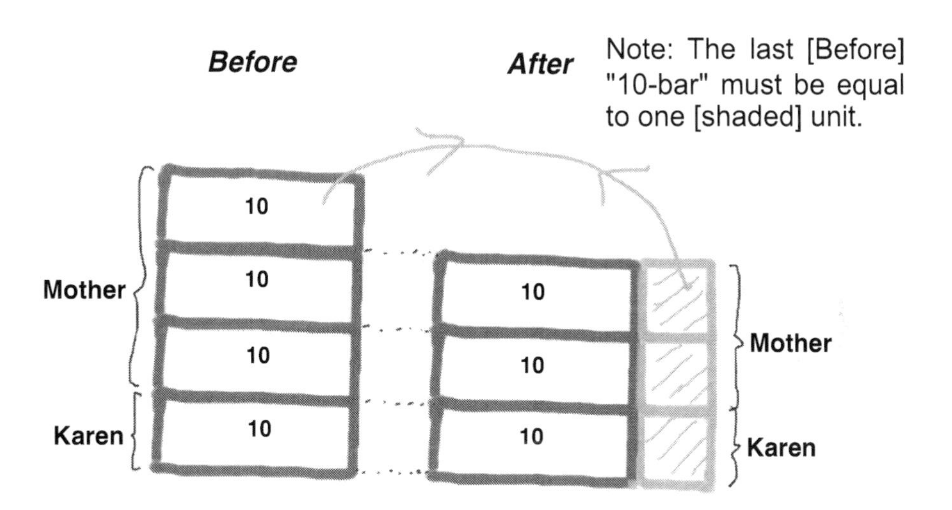

From the model,

1 [shaded unit] = 10

Karen's mother will be twice as old as her 10 years later.

Note: In solving this question, we never make use of the fact that the age difference between mother and daughter remains unchanged at any point in time. However, this useful fact will be applied in the next worked example.

Thought Process

Karen is 10 years old and her mother is 30 years old. How many years later will Karen's mother be twice as old as her?

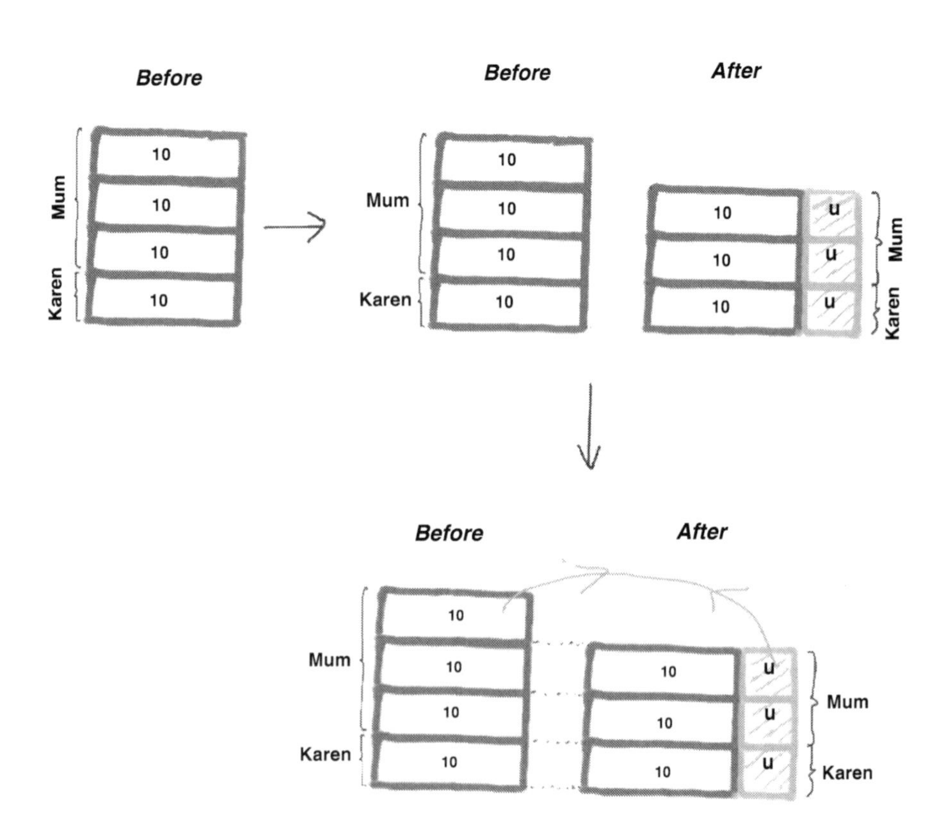

Thought Process

Karen is 10 years old and her mother is 30 years old. How many years later will Karen's mother be twice as old as her?

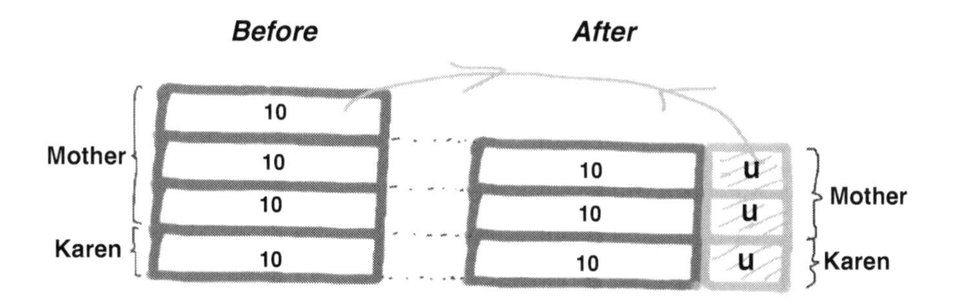

Observe:

	Before	*After*
Karen	10	10 + u
Mother	30	10 + 10 + u + u

From the mother's age, out of the "10 + 10 + u + u", one "u" must represent the number of years later when she will be twice as old as Karen. This means that the remaining terms, "10 + 10 + u", must represent her age before, which is 30.

So, 10 + 10 + u = 30
u = 30 – 10 – 10 = 10

In other words, in 10 years, Karen's mother will be twice as old as her.

Worked Example 18

When Sally was 6 years old, her mother was 34 years old. If her mother is 3 times as old as Sally now, what is her mother's present age?

Method 1

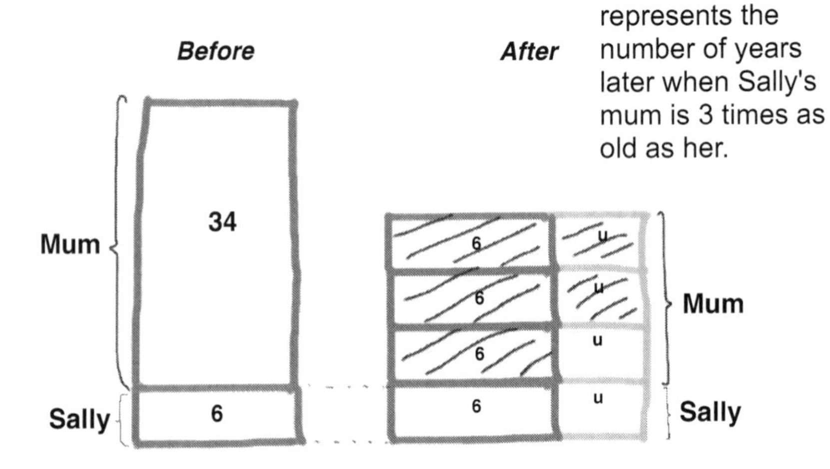

From the model,

6 + 6 + 6 + 2 units → 34
18 + 2 units → 34
2 units → 34 – 18 = 16
1 unit → 16 ÷ 2 = 8
34 + 1 unit → 34 + 8 = 42

Sally's mother is 42 years old now.

Check:
34 + 8 = 42; 6 + 8 = 14
42 = 3 × 14

Note: In this method, we don't make use of the fact that the age difference between mother and daughter remains unchanged at any point in time.

Thought Process

Method 1

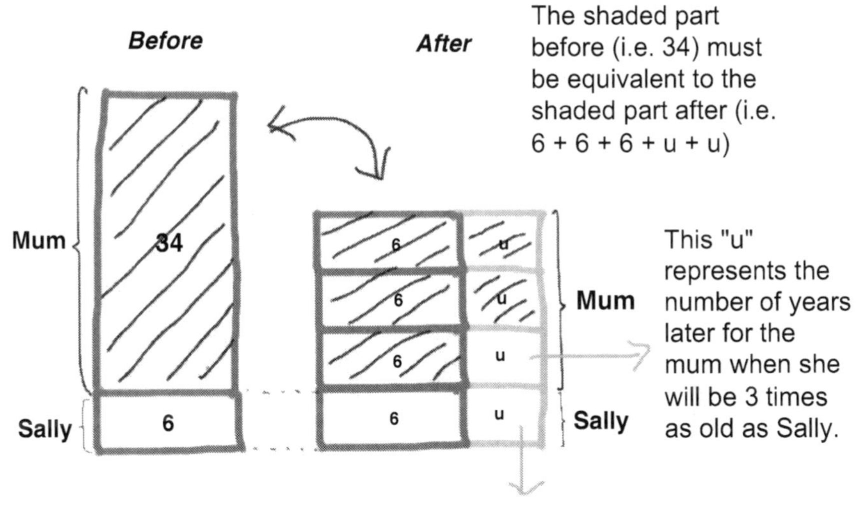

From the model,

This "u" represents the number of years later for Sally when her mum will be 3 times as old as her.

$6 + 6 + 6 + 2u \rightarrow 34$
$18 + 2u \rightarrow 34$
$2u \rightarrow 34 - 18 = 16$
$u \rightarrow 16 \div 2 = 8$
$34 + u \rightarrow 34 + 8 = 42$

Sally's mother is 42 years old now.

When Sally was 6 years old, her mother was 34 years old. If her mother is 3 times as old as Sally now, what is her mother's present age?

Method 2

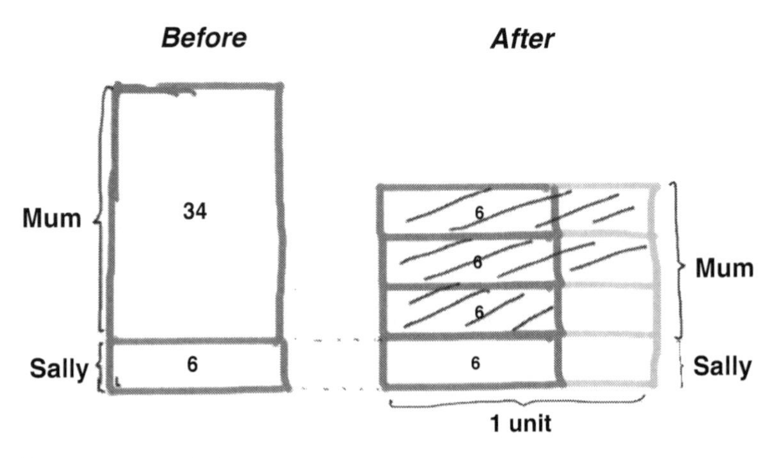

From the model,

2 units + 6 → 34
2 units → 34 – 6 = 28 1
unit → 28 ÷ 2 = 14
1 unit – 6 → 14 – 6 = 8

Sally's mother will be 3 times as old as her in 8 years' time.

34 + 8 = 42
Sally's present age is 42 years old.

Note: Unlike in Method 1, "1 unit" here is defined as the age of Sally when her mum will be 3 times as old as her, as compared to the number of years when this event will take place.

No prior knowledge that the age difference remains unchanged is needed.

Thought Process

Method 2

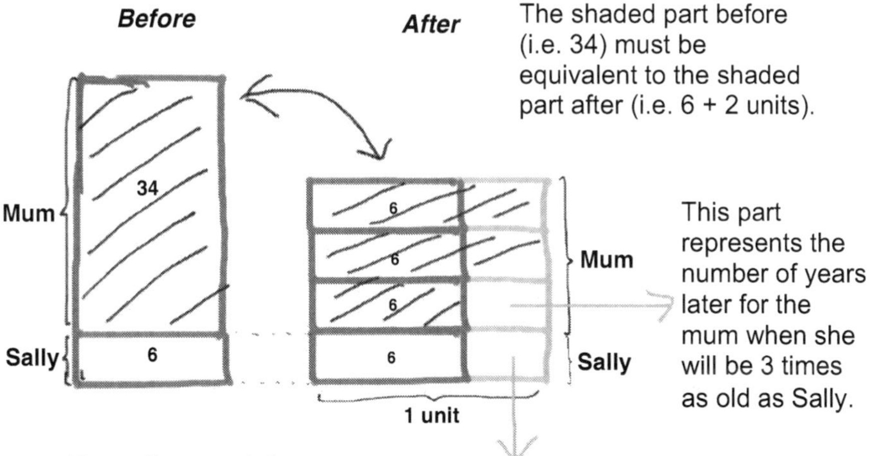

From the model,
2 units + 6 → 34
2 units → 34 – 6 = 28
1 unit → 28 ÷ 2 = 14
1 unit – 6 → 14 – 6 = 8

Sally's mother would be 3 times as old as her in 8 years' time.

34 + 8 = 42
Sally's present age is 42 years old.

When Sally was 6 years old, her mother was 34 years old. If her mother is 3 times as old as Sally now, what is her mother's present age?

Method 3

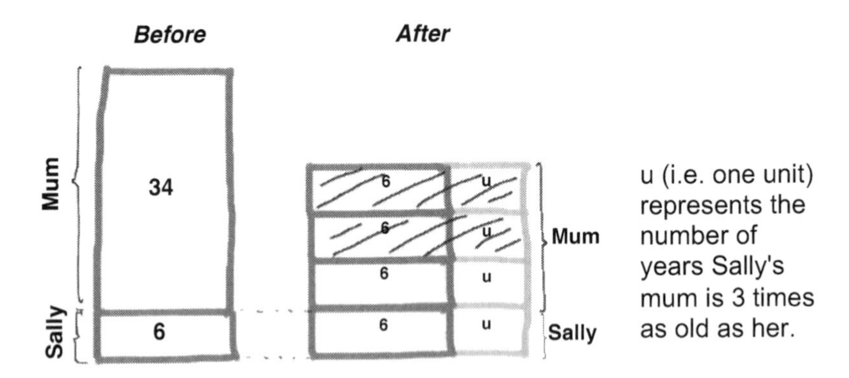

The age difference between them does not change at any time.

From the model,

6 + 6 + 2 units → 34 – 6 = 28
12 + 2 units → 28
2 units → 28 – 12 = 16
1 unit → 16 ÷ 2 = 8
34 + 1 unit → 34 + 8 = 42

Sally's mother is 42 years old now.

Thought Process

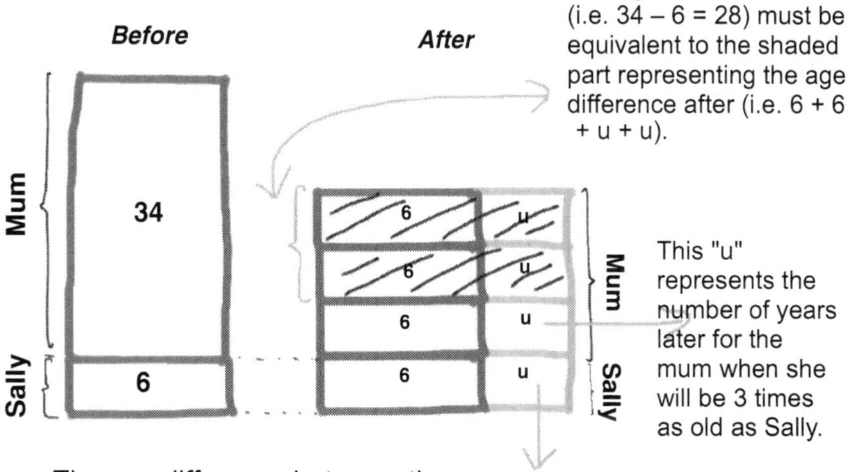

This "u" represents the number of years later for Sally when her mum will be 3 times as old as her.

The age difference between them does not change at any time, which is 34 – 6 = 28.

From the model,

$6 + 6 + 2u \rightarrow 34 - 6 = 28$
$12 + 2u \rightarrow 28$
$2u \rightarrow 28 - 12 = 16$
$u \rightarrow 16 \div 2 = 8$
$34 + u \rightarrow 34 + 8 = 42$

Sally's mother is 42 years old now.

When Sally was 6 years old, her mother was 34 years old. If her mother is 3 times as old as Sally now, what is her mother's present age?

Method 4

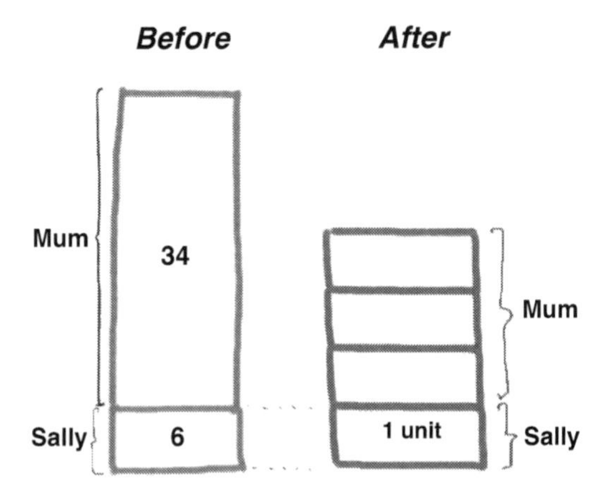

The age difference between them does not change at any time, which is 34 – 6 = 28.

From the model,
2 units → 28
1 unit → 28 ÷ 2 = 14
3 units → 3 × 14 = 42

Sally's mother is 42 years old now.

Note: Both Methods 3 and 4 make use of the fact that the age difference does not change.

In Method 3, 1 unit is defined as the number of years Sally's mum is 3 times as old as her.

In Method 4, 1 unit is defined as Sally's present age.

Thought Process

Method 4

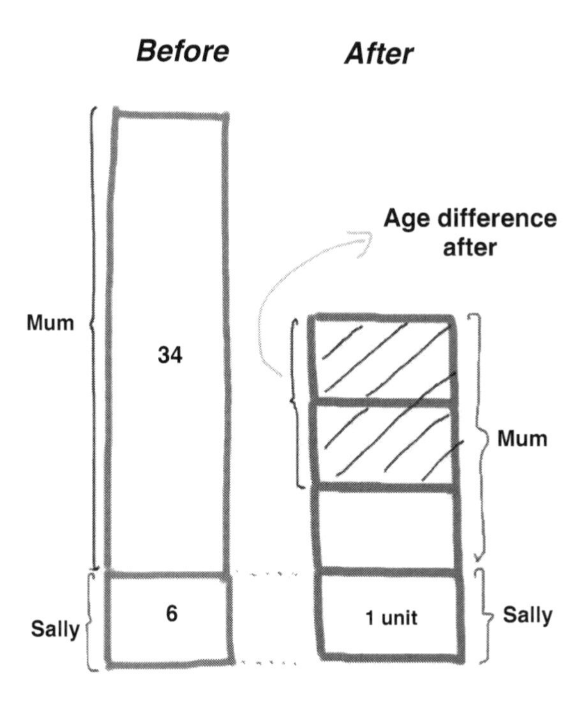

The age difference between them does not change at any time, which is 34 – 6 = 28.

Age difference after = age difference before
2 units → 28
1 unit → 28 ÷ 2 = 14
3 units → 3 × 14 = 42

Sally's mother is 42 years old now.

Th!nk CRE8TIVELY

Both Methods 3 and 4 depend on the fact that the age difference does not change at any point in time.

In Method 3, "1 unit" represents the number of years Sally's mum is 3 times as old as her.

In Method 4, "1 unit" represents Sally's present age.

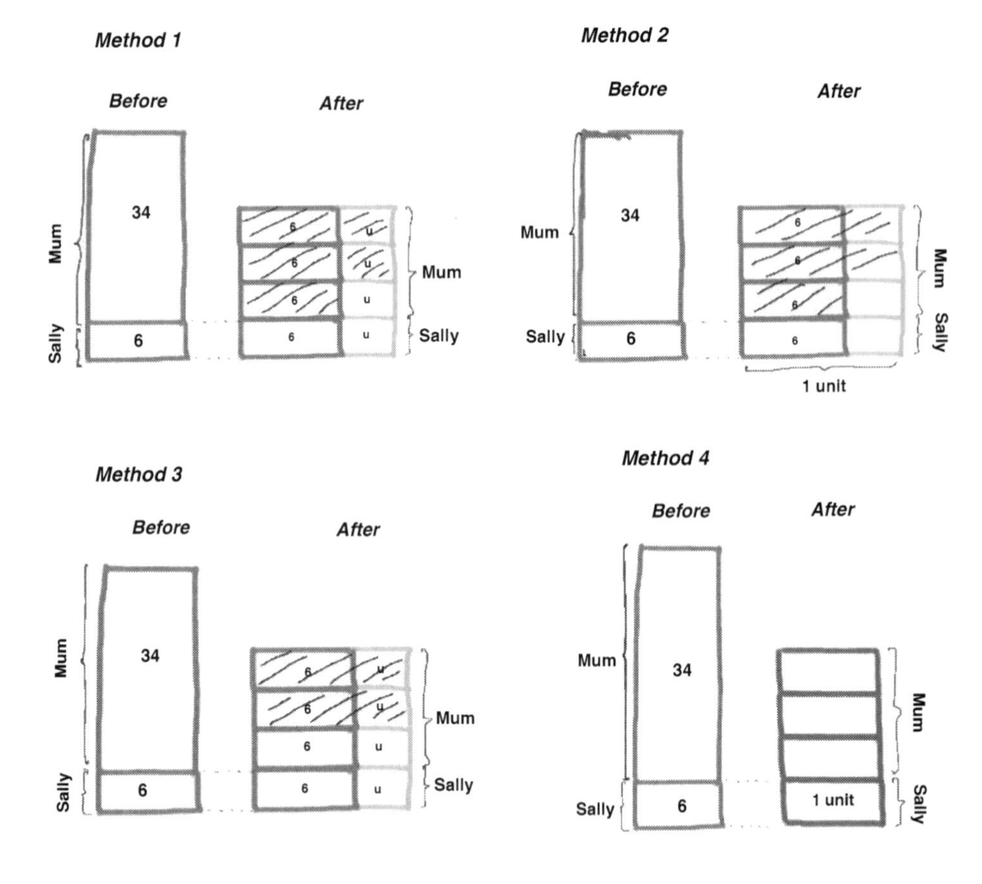

Thought Process

Methods 1–4

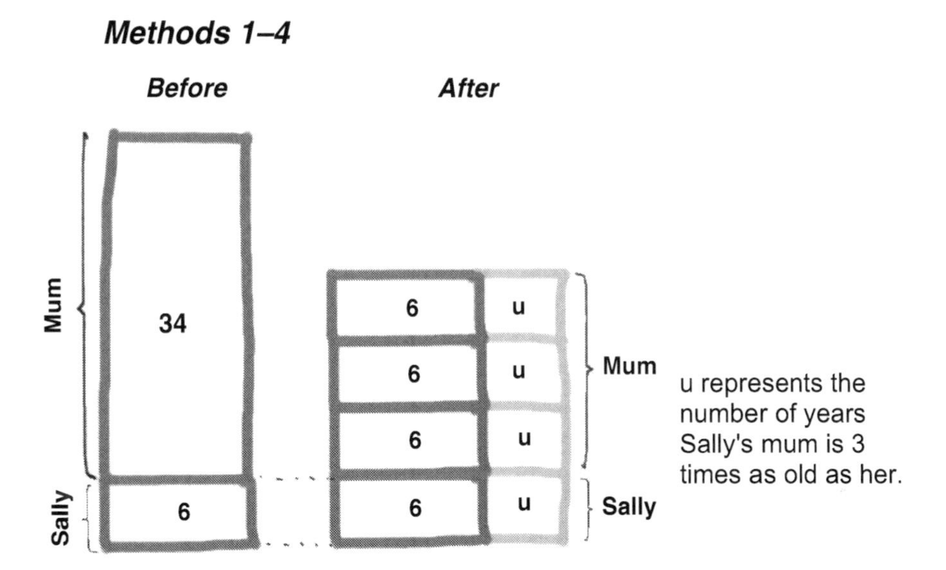

Observe that in methods 1 and 3, we make use of the same stack model drawing; however, the thinking or reasoning process is different in each case. Method 3 makes use of the fact that the age difference remains unchanged at any point in time.

Similarly, methods 2 and 3 share the same stack model drawing, but the thought process differs in each case.

In methods 1 and 3, one unit (or u) represents the number of years Sally's mum is 3 times as old as her. In methods 2 and 4, one unit is defined as Sally's present age.

In methods 1 and 2, we don't assume any prior knowledge that the age difference between mother and daughter remains unchanged at any point of time. It is up to us to define what one unit is, which then translates into a different equation.

Worked Example 19

Laval is 18 years older than Chris. How old will Chris be when Laval is 3 times as old as Chris?

Method 1

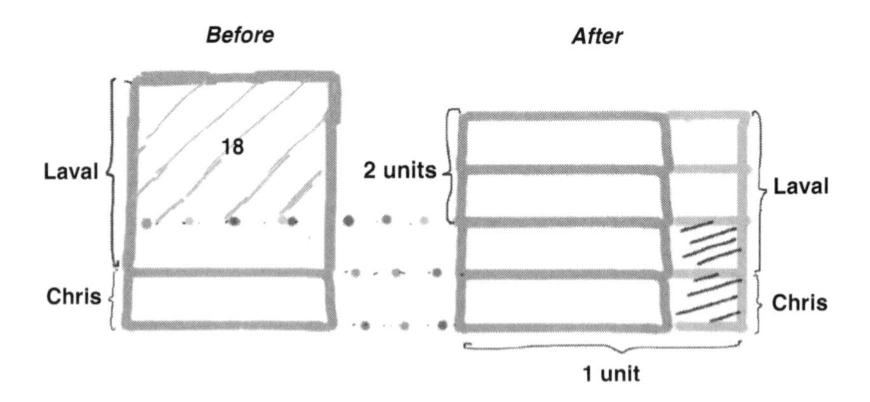

From the model,

2 units → 18
1 unit → 9

Chris will be 9 years old when Laval will be 3 times as old as her.

TH!NK

In how many years' time will Laval be 3 times as old as Chris?

Thought Process

Laval is 18 years older than Chris. How old will Chris be when Laval is 3 times as old as Chris?

Method 1

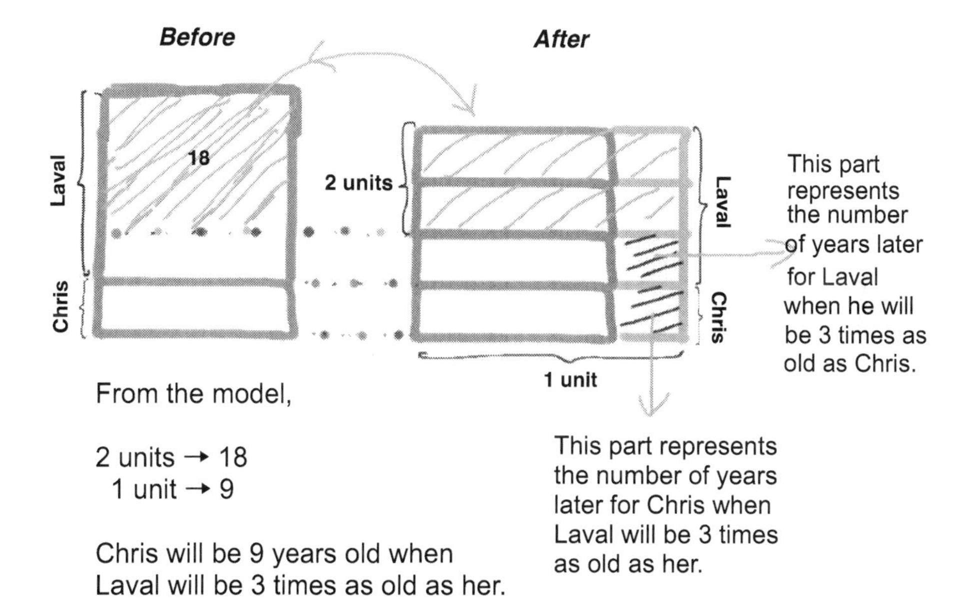

From the model,

2 units → 18
1 unit → 9

Chris will be 9 years old when Laval will be 3 times as old as her.

Note that in this method, we do not make use of the fact that the age difference between Chris and Laval does not change at any point in time, although it may appear that we make use of this fact. The relationship is obtained only by comparing the *before* and *after* models.

Same, yet different

TH!NK

Observe that although we can find out the age of Chris when Laval will be three times as old as her, however, we do not know in how many years' time this will take place.

In this question, we are not given the age of Chris and of Laval, but only the difference in their ages, as compared to the previous question involving Sally and her mum, when the age of each person was given.

In the previous question, we were able to find the number of years later when Sally's mum will be three times as old as her, and also how old the mother will be by then.

On the surface, both age-related questions look similar, but the second question is conceptually more interesting than the first one. *Do you know in how many years' time will Laval be 3 times as old as Chris?*

Laval is 18 years older than Chris. How old will Chris be when Laval is 3 times as old as Chris?

Method 2

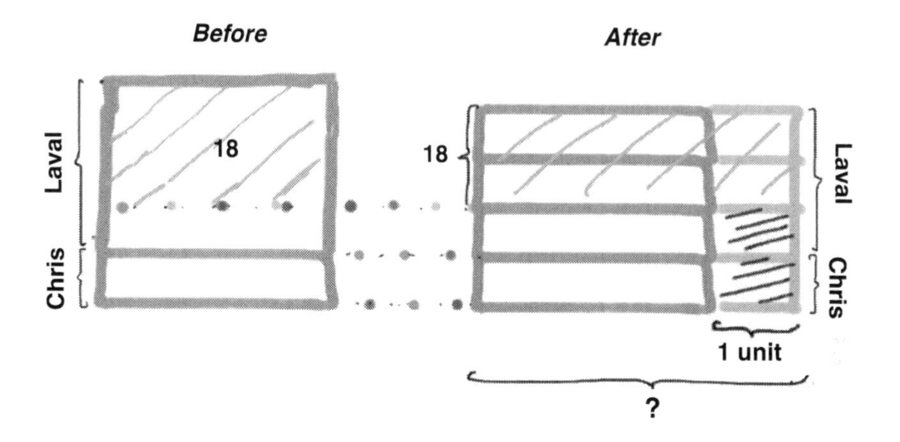

From the model, we cannot determine the value of 1 unit; however, we can find Chris's age, which is given by $18 \div 2 = 9$.

Chris will be 9 years old when Laval will be 3 times as old as her.

TH!NK

In how many years' time will Laval be 3 times as old as Chris?

Thought Process

Laval is 18 years older than Chris. How old will Chris be when Laval is 3 times as old as Chris?

Method 2

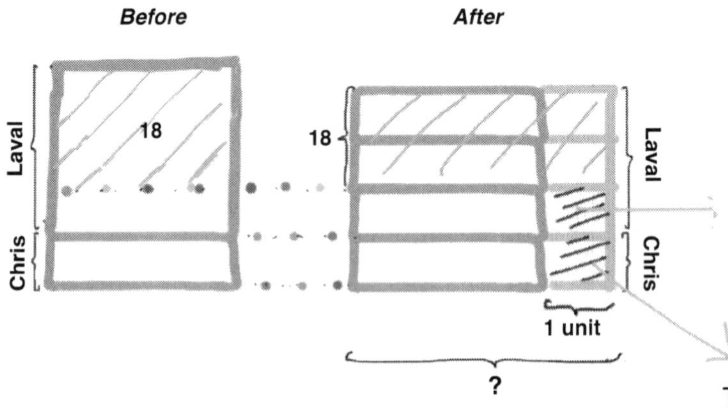

This part represents the number of years later for Laval, when he will be 3 times as old as Chris.

This part represents the number of years later for Chris, when Laval will be 3 times as old as her.

In this example, we define one unit to be the number of years later when Laval will be three times as old as Chris.

As the model drawing shows, we cannot determine the value of 1 unit; however, by comparing the parts of the before and after models, using the dotted lines, we can find Chris's age, which is given by $18 \div 2 = 9$.

Thus, Chris will be 9 years old when Laval will be 3 times as old as her.

Note that as in Method 1, we did not use the fact that the age difference between Chris and Laval doesn't change at any point in time.

Laval is 18 years older than Chris. How old will Chris be when Laval is 3 times as old as Chris?

Method 3

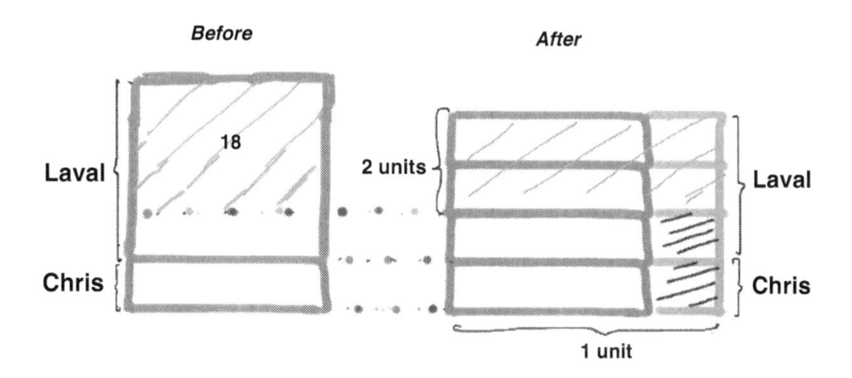

The age difference of 18 years between Chris and Laval does not change at any point in time.

From the model,

2 units → 18
1 unit → 9

Chris will be 9 years old.

Note: The computations for Methods 1 and 3 look the same, but the thought processes are different. Unlike Method 3 that makes use of the fact that the age difference does not change, Method 1 does not.

Thought Process

Laval is 18 years older than Chris. How old will Chris be when Laval is 3 times as old as Chris?

Method 3

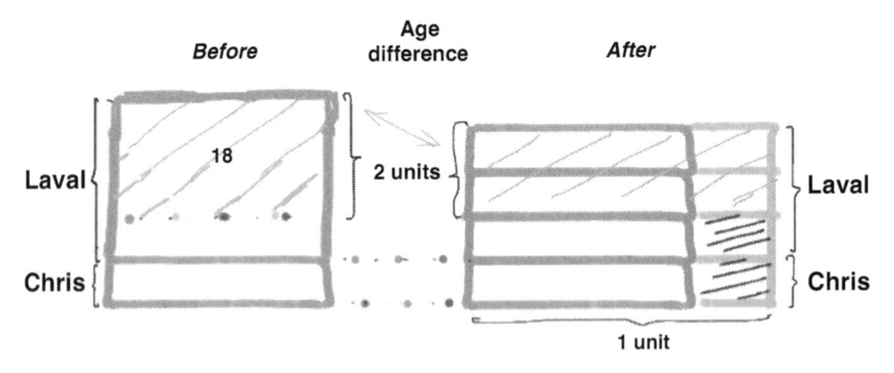

From the model,

Age difference after = Age difference before

2 units → 18

1 unit → 9

Chris will be 9 years old.

Note: Using the fact that the age difference does not change, had we defined one unit to be the number of years later when Laval will be 3 times as old as Chris, we would not be able to find out the answer. However, that does not in any way prevent us from finding the age of Chris, when Laval will be 3 times as old as her.

Worked Example 20

In 2012, Sam's grandfather was 4 times as old as him. In 2020, the grandfather will be 3 times as old as him. How old will the grandfather be in 2025?

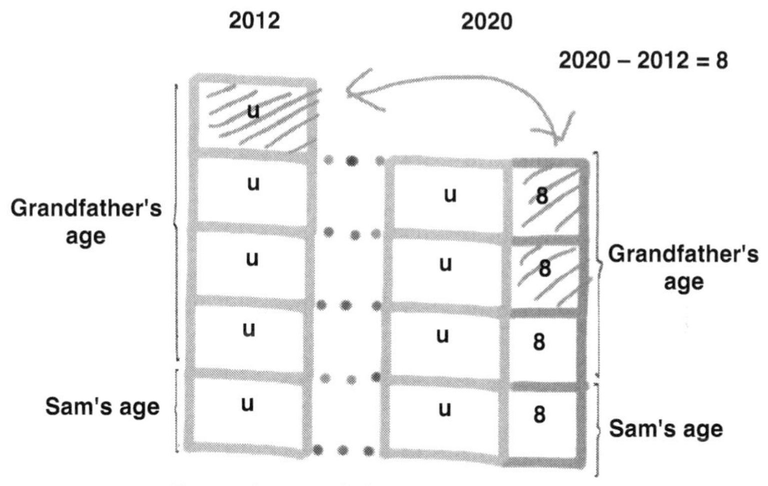

From the model,

u → 8 + 8 = 16
4u → 4 × 16 = 64
Sam's grandfather was 64 years old in 2012.

2025 – 2012 = 13
64 + 13 = 77
Sam's grandfather will be 77 years old in 2025.

Practice

In 2000, Jane's father was 5 times as old as her. In 2012, he was twice as old as her. How old will Jane's father be in 2020? Answer: 40 years old.

Thought Process

In 2012, Sam's grandfather was 4 times as old as him. In 2020, the grandfather will be 3 times as old as him. How old will the grandfather be in 2025?

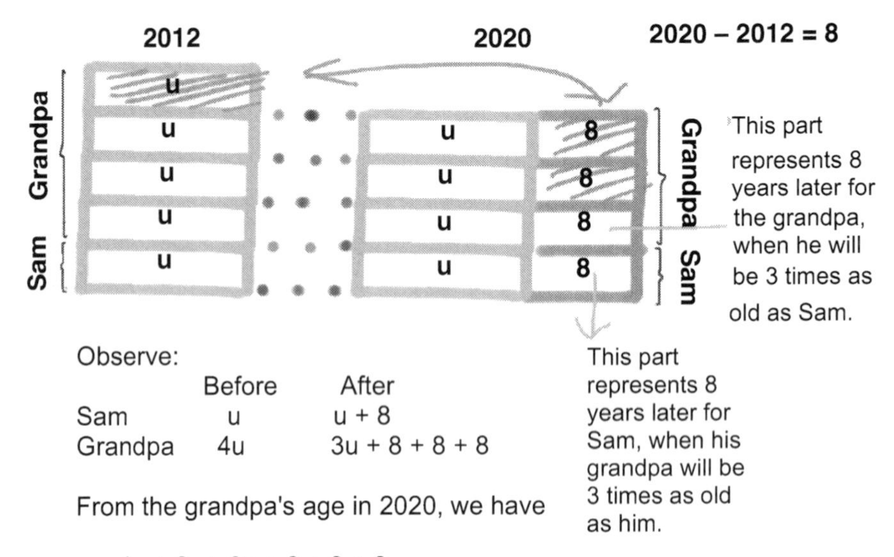

Observe:

	Before	After
Sam	u	u + 8
Grandpa	4u	3u + 8 + 8 + 8

From the grandpa's age in 2020, we have

$$4u + 8 \rightarrow 3u + 8 + 8 + 8$$
$$3u + u + 8 \rightarrow 3u + 8 + 8 + 8$$
$$u \rightarrow 8 + 8 = 16$$

Note that we may also obtain the above result of "u → 8 + 8 = 16" by comparing the *before* and *after* models, as depicted by the shaded parts.

So, in 2012, Sam was 16 years old, and his grandpa was 4 × 16 = 64 years old.

From 2012 to 2025, there are 13 years.

So, in 2025, Sam's grandpa will be 64 + 13 = 77 years old.

Simultaneous Equations for Kids

Simultaneous Equations for Kids — Examples 21–24

Traditionally, questions on simultaneous equations, or word problems involving these linear equations, are read in secondary 1–2 in Singapore. However, the visualisation nature of bar- and of stack-modelling has enabled these secondary maths (or middle-school) questions to be posed in lower grades.

Unlike bar modelling, stack modelling often allows one to form different meaningful relationships between the two variables, because one can always stack the bars or units up and down, besides stacking them in a left or right manner. Because of this stacking flexibility, stack modelling is like 2D modelling, as compared to bar modelling, which tends to be rigidly practised in one dimension.

Moreover, stack modelling, when applied to solving simultaneous equations, tends to promote creative thinking, as one tries to figure out a number of interesting numerical relationships between the variables.

Worked Example 21

One mango and one pear cost $3.30.
Two mangoes and five pears cost $9.00.
How much does a mango cost?

Method 1

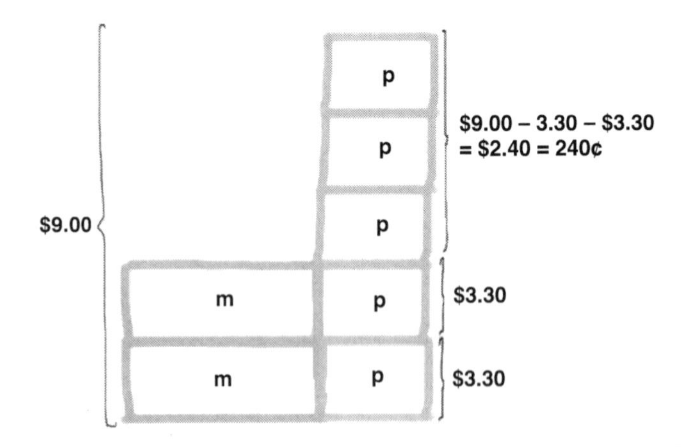

From the model,

p → 240¢ ÷ 3 = 80¢
m + p → 330¢ (given)
m → 330¢ – 80¢ = 250¢

A mango costs $2.50.

Unlike in bar modelling, where models tend to be drawn from left to right, or right to left, in stack modelling, models can be drawn more flexibly, as one can go from to bottom to top, and left to right.

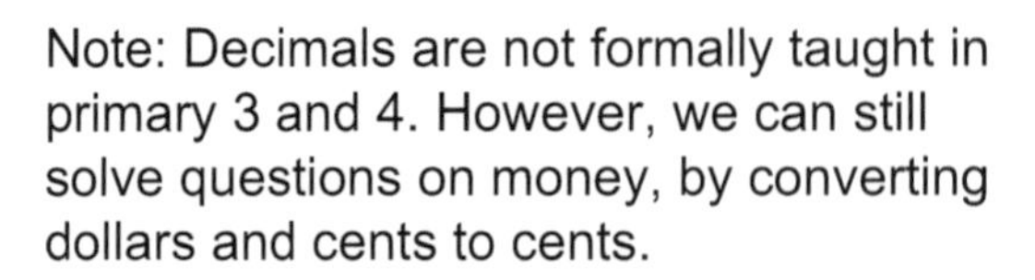

Note: Decimals are not formally taught in primary 3 and 4. However, we can still solve questions on money, by converting dollars and cents to cents.

Thought Process

One mango and one pear cost \$3.30.
Two mangoes and five pears cost \$9.00.
How much does a mango cost?

Instead of adding 2 more mangoes and 5 pears on top on the 1 mango and 1 pear, already drawn, we only add in 1 more mango and 4 more pears to match the second statement.

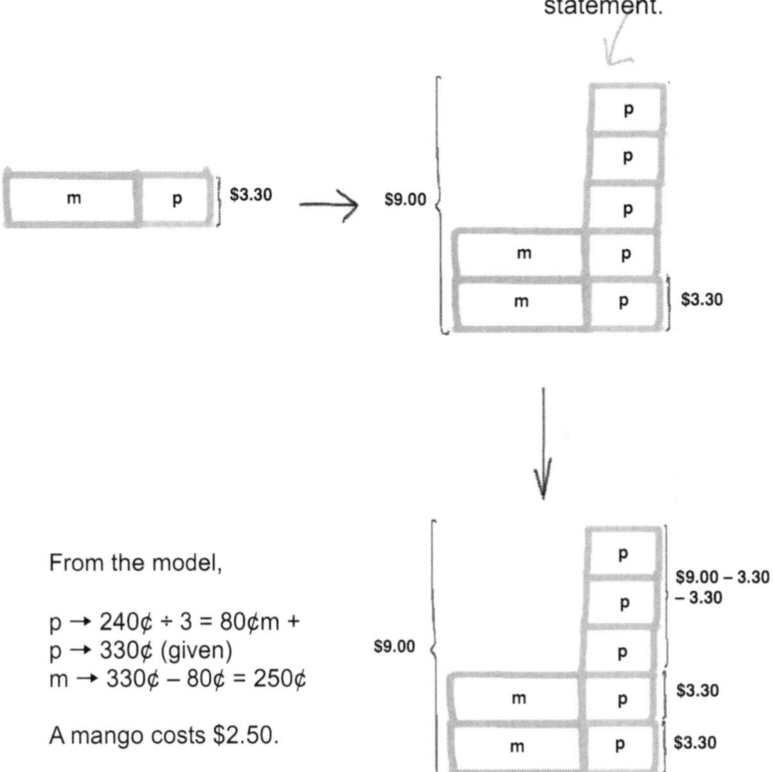

From the model,

p → 240¢ ÷ 3 = 80¢m +
p → 330¢ (given)
m → 330¢ – 80¢ = 250¢

A mango costs \$2.50.

One mango and one pear cost $3.30.
Two mangoes and five pears cost $9.00.
How much does a mango cost?

Method 2

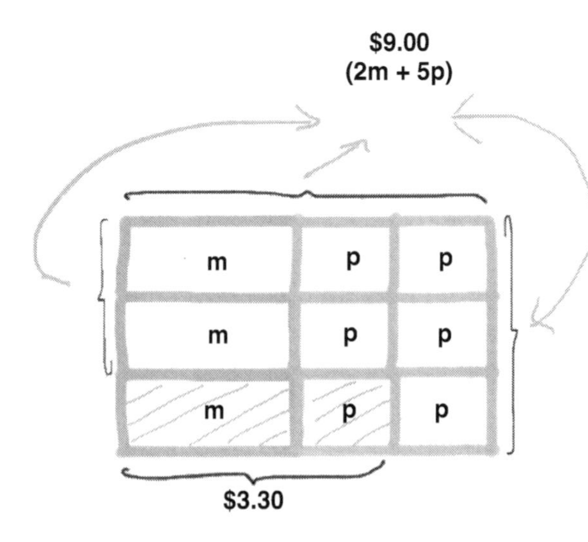

Bar-modelling is like one-dimensional modelling; stack-modelling is like two-dimensional modelling.

From the model,

$9.00 + $3.30 = 900¢ + 330¢ = 1230¢
m + 2p → 1230¢ ÷ 3 = 410¢

m + p → 330¢ (given)
2m + 2p → 330¢ + 330¢ = 660¢
m → 660¢ – 410¢ = 250¢ = $2.50

or m + p → 330¢ (given)
p → 410¢ – 330¢ = 80¢
m → 330¢ – 80¢ = 250¢

A mango costs $2.50.

Thought Process

One mango and one pear cost $3.30.
Two mangoes and five pears cost $9.00.
How much does a mango cost?

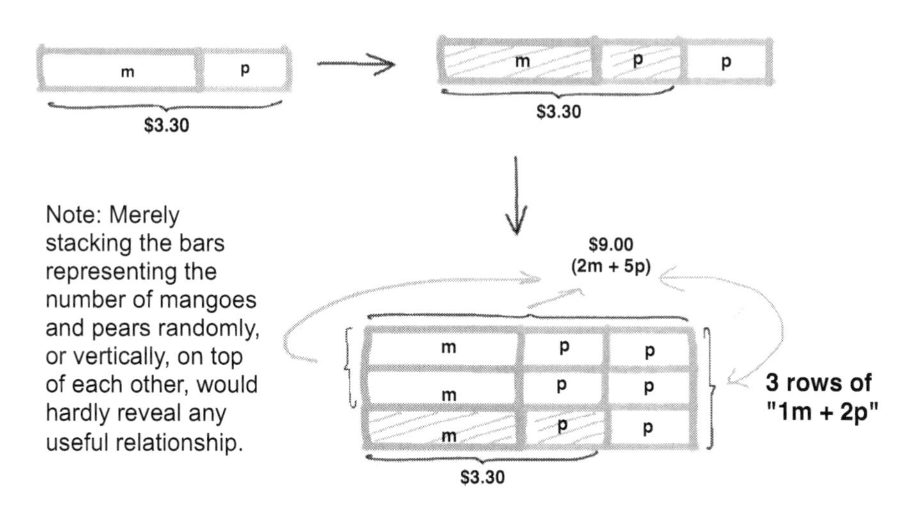

From the model,

3 rows of "m + 2p" → $9.00 + $3.30
= 900¢ + 330¢ = 1230¢

1 row of "m + 2p" → 1230¢ ÷ 3
= 410¢

m + p → 330¢ (given)
2m + 2p → 330¢ + 330¢ = 660¢
m → 660¢ – 410¢ = 250¢ = $2.50

Alternatively,
m + p → 330¢ (given)
p → 410¢ – 330¢ = 80¢
m → 330¢ – 80¢ = 250¢

A mango costs $2.50.

Worked Example 22

One orange and one apple cost $1.05. Two oranges and three apples cost $2.75. How much do two apples cost?

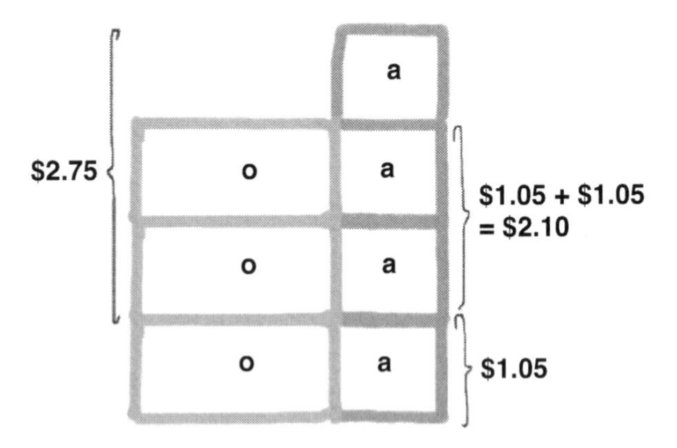

From the model,

a → $2.75 – $2.10 = $0.65
2a → 2 × $0.65 = $1.30

Two apples cost $1.30.

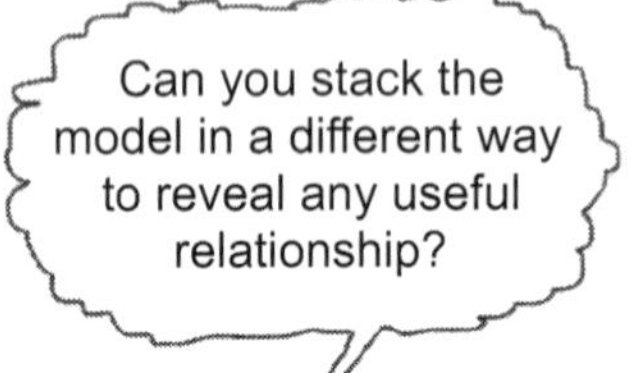

Or, 3o + 3a → 3 × $1.05 = $3.15
2o + 3a → $2.75
o → $3.15 – $2.75 = $0.40
a → $1.05 – $0.40 = $0.65
2a → 2 × $0.65 = $1.30
Two apples cost $1.30.

Practice

Two similar e-books and three similar apps cost $30. One such e-book and two such apps cost $17. What is the cost of one e-book? Answer: $9.

Worked Example 23

One pear and two oranges cost \$1.00.
Two pears and one orange cost \$1.10.
What is the total cost of two pears and two oranges?

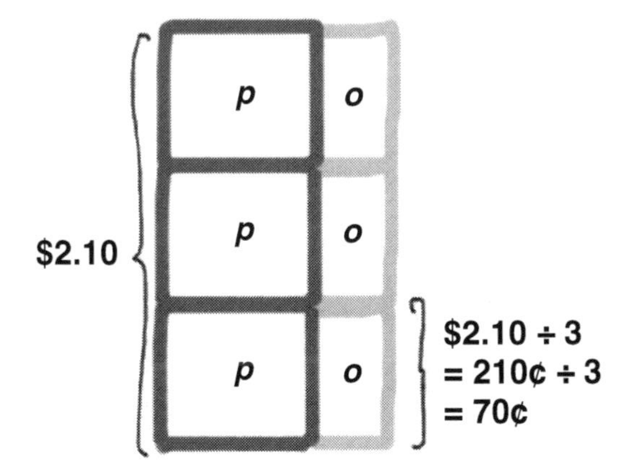

From the model,

$\$1.00 + \$1.10 = \$2.10$; $p + o \rightarrow 70¢$
$2p + 2o \rightarrow 70¢ + 70¢ = 140¢ = \$1.40$

Two pears and two oranges cost \$1.40.

Practice

Two smartphones and three tablets cost \$910.
Two tablets and three smartphones cost \$840.
What is the cost of two smartphones and two tablets? Answer: \$700.

Thought Process

One pear and two oranges cost $1.00.
Two pears and one orange cost $1.10.
What is the total cost of two pears and two oranges?

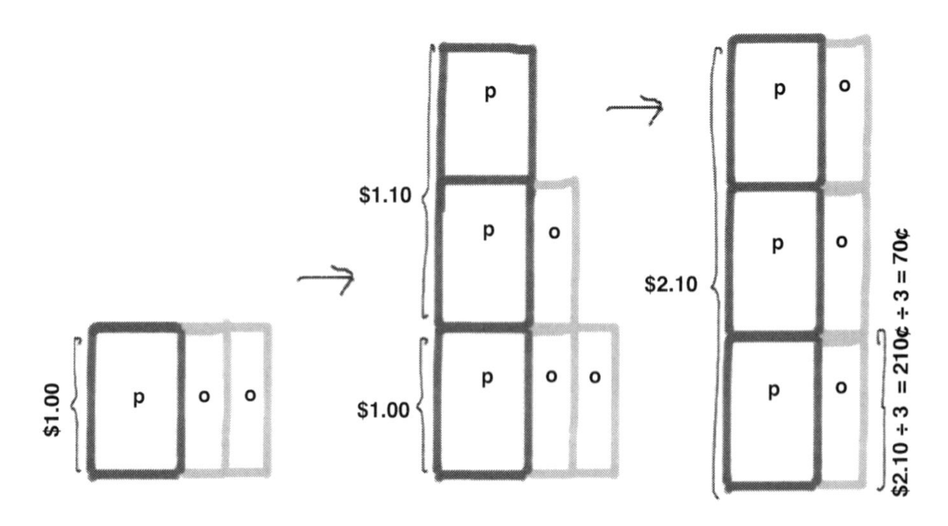

From the model,

3 rows of "p + o" → 210¢
1 row of "p + o" → 70¢
2 rows of "p + o" → 70¢ + 70¢ = 140¢ = $1.40

Two pears and two oranges cost $1.40.

Worked Example 24

Six pears and five oranges cost \$8.70. Five such pears and four such oranges cost \$1.60 less. What is the cost of three oranges?

Method 1

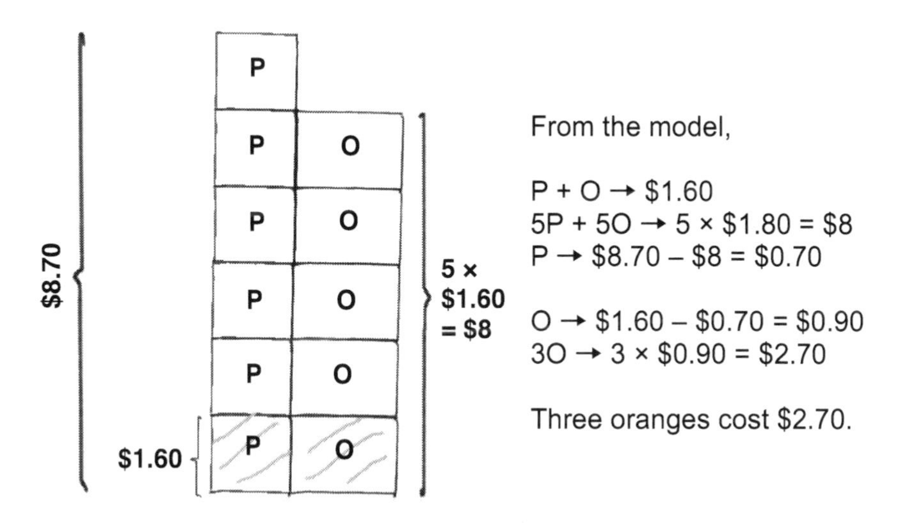

From the model,

P + O → \$1.60
5P + 5O → 5 × \$1.80 = \$8
P → \$8.70 – \$8 = \$0.70

O → \$1.60 – \$0.70 = \$0.90
3O → 3 × \$0.90 = \$2.70

Three oranges cost \$2.70.

6 pears + 5 oranges cost \$8.70. (given)
5 pears + 4 oranges cost \$1.60 less. (given)

This means that (6 – 5) = 1 pear plus (5 – 4) = 1 orange cost \$1.60.

1 pear + 1 orange → \$1.60

Worked Example 24

Six pears and five oranges cost \$8.70. Five such pears and four such oranges cost \$1.60 less. What is the cost of three oranges?

Method 2

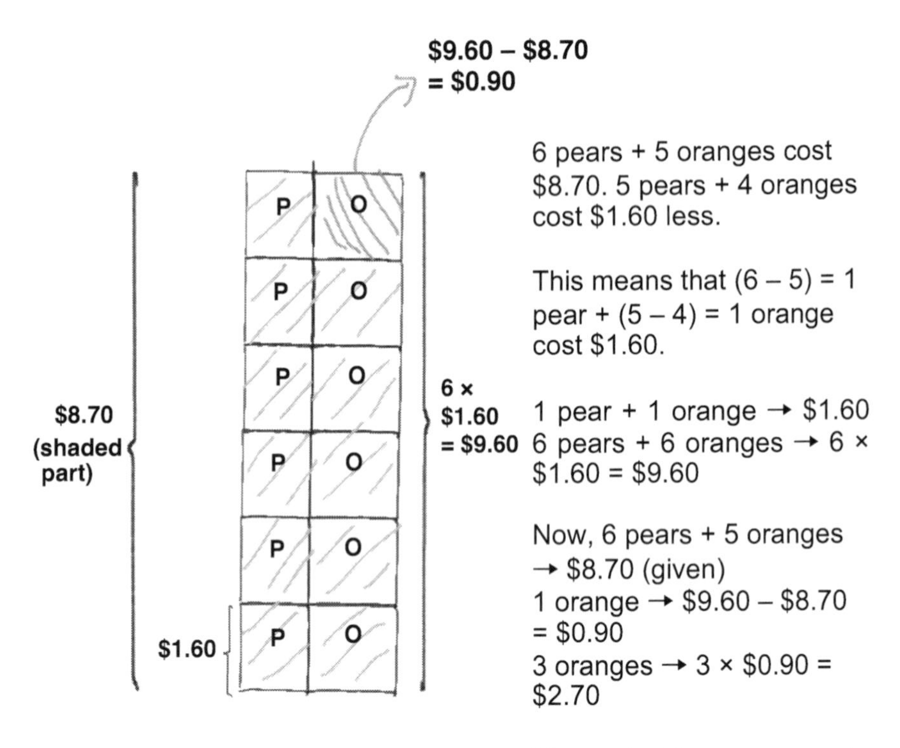

6 pears + 5 oranges cost \$8.70. 5 pears + 4 oranges cost \$1.60 less.

This means that (6 – 5) = 1 pear + (5 – 4) = 1 orange cost \$1.60.

1 pear + 1 orange → \$1.60
6 pears + 6 oranges → 6 × \$1.60 = \$9.60

Now, 6 pears + 5 oranges → \$8.70 (given)
1 orange → \$9.60 – \$8.70 = \$0.90
3 oranges → 3 × \$0.90 = \$2.70

The cost of three oranges is \$2.70.

Length

Length — Worked Examples 25–28

Like those on *Whole Numbers*, solving questions on *Length*, which use the stack model method, often offers an advantage over the bar model method. This is because comparison may be done using only one model drawing, instead of constructing different bar models, then comparing them to derive any meaningful numerical relationship.

Worked Example 25

The total length of string P and string Q is 818 cm. String P is 342 cm shorter than string Q. What is the length of string Q in metres and centimetres?

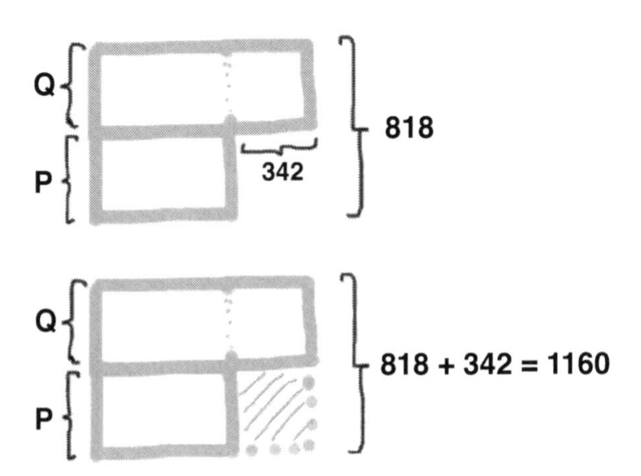

From the model,

818 + 342 = 1160
1160 ÷ 2 = 580

String Q is 580 cm long.

Note: A longer method would be to equate the length of string P to one unit, then to find the value of one unit, before assigning the value of string Q to "one unit + 342".

Practice

Two pieces of ribbon measure a total length of 938 cm. One ribbon is 280 cm longer than the other. What is the length of the longer ribbon in metres and centimetres?

Answer: 6 m 9 cm.

Worked Example 26

Sam and Jim have a total height of 327 cm. If Sam is 180 cm tall, how much taller is he than Jim?

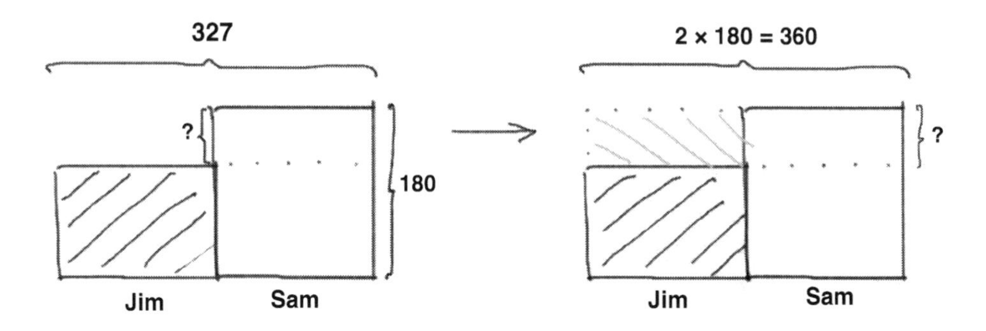

From the model,

2 × 180 = 360
360 – 327 = 33

Sam is 33 cm taller than Jim.

Note: We needn't find the height of Jim first in order to find how much taller Sam is than Jim.

Worked Example 27

String A is twice as long as string B. String C is three times as long as string B. The total length of the three strings is 624 cm. What is the length (in metres and centimetres) of string A?

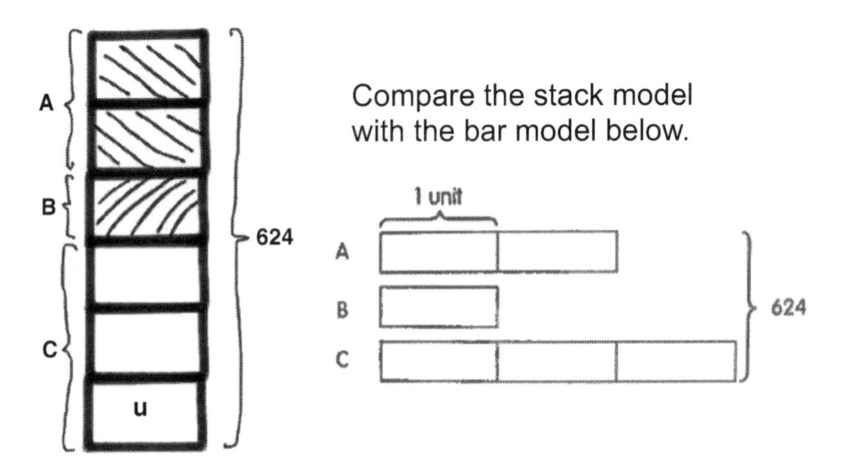

From the model,

6u → 624
u → 624 ÷ 6 = 104
2u → 2 × 104 = 208 or 2u → 624 ÷ 3 = 208

208 cm = 2 m 8 cm
String A is 2 m 8 cm long.

Practice

Three boxes, P, Q and R, weigh a total of 102 kg. Box Q weighs twice as heavy as box P, and box R is three times as heavy as box P. What is the mass of box R? Answer: 51 kg.

Worked Example 28

Dave and Joy have a total height of 265 cm. Joy and Rick have a total height of 254 cm. If Rick is 135 cm tall, what is the height of Dave? Express your answer in metres and centimetres.

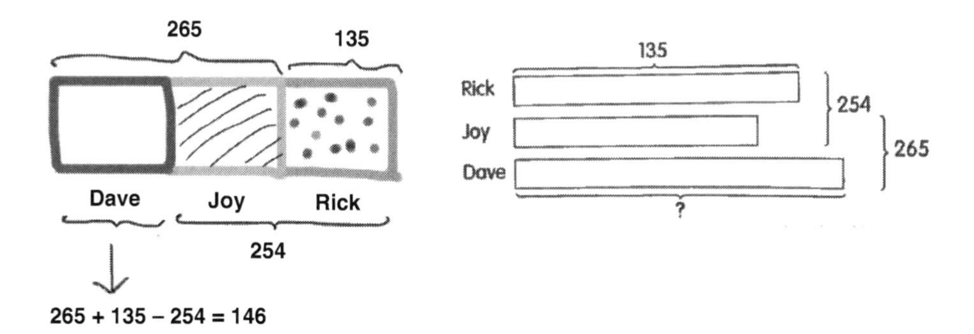

From the model,

Dave + Joy + Rick → 265 + 135
Dave → 265 + 135 – 254 = 146

Dave is 1 m 46 cm tall.

Note: Unlike in bar modelling, where we need to know whether a bar is longer or shorter than another bar, in stack modelling, such concern does not arise.

Practice

The total length of Lisa and Claire is 280 cm. The total length of Claire and Yvonne is 295 cm. If Yvonne's height is 145 cm, how tall is Lisa? Express your answer in metres and centimetres.

Answer: 1 m 30 cm.

Thought Process

Dave and Joy have a total height of 265 cm. Joy and Rick have a total height of 254 cm. If Rick is 135 cm tall, what is the height of Dave? Express your answer in metres and centimetres.

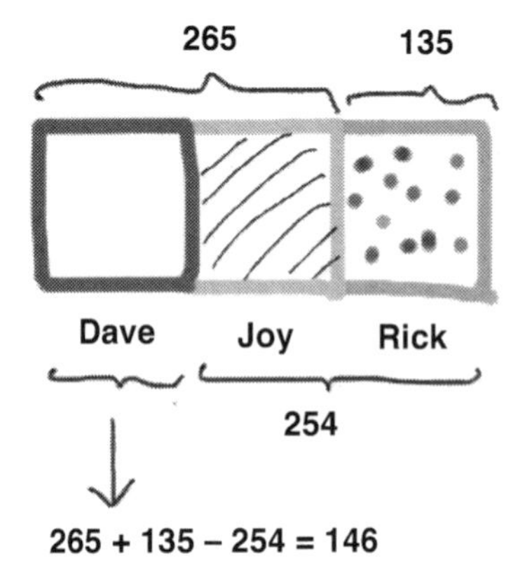

Dave + Joy + Rick → 265 + 135
Dave → 265 + 135 – 254 = 146

Dave is 1 m 46 cm tall.

Stacking the bars next to each other enables us to find the answer in one step, as compared to the bar model method, which requires a two-step process.

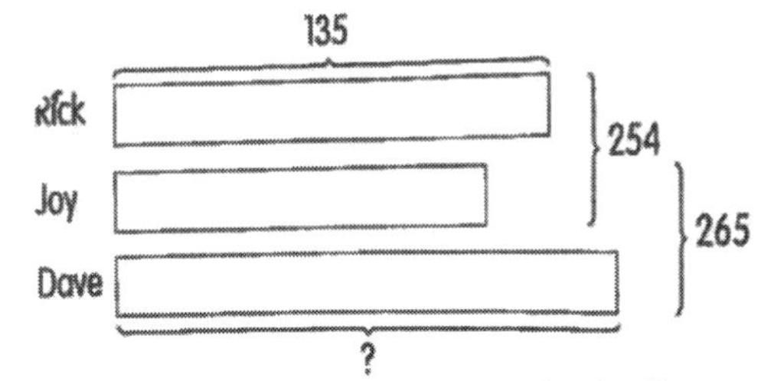

In comparing three quantities, especially when at least two of them are close to each other, in practice, it is not easy to draw the three bars sensibly proportionate, so that the three bars are depicted correctly.

Often times, we already know which bar is longer or shorter than another bar, because we work backwards to determine their respective lengths, before drawing the bars — this is like finding the answer first before depicting the bar models.

The way bar models are generally constructed often blinds us from seeing numerical relationships, obvious as they may be; however, because stack models offer us the flexibility to represent the bars in two dimensions, it is not uncommon to find the answer more effectively, for example, in finding the answer in fewer steps.

Before-After Word Problems

Before-After Word Problems — Examples 29–34

Before-After word problems are the types of non-routine questions that are probably most favourable to the stack model method. The examples compare the stack method *vis-à-vis* other problem-solving heuristics commonly used by local teachers in Singapore, such as the bar model and the unitary methods.

In general, these before-after questions are traditionally solved using algebra in secondary school, but thanks to bar- and stack-modelling, these word problems can now be given to primary 3–4 above-average pupils to enhance their mathematical problem-solving skills. Besides, encouraging pupils to come up with different methods of solution (related to different stack models), based on a particular problem-situation, motivates them to think creatively and intuitively.

Worked Example 29

There were twice as many women as men in a hall. After 350 women left and 450 men went in, there were 3 times as many men as women. If 1000 adults remained, how many adults were there in the hall at first?

Method 1

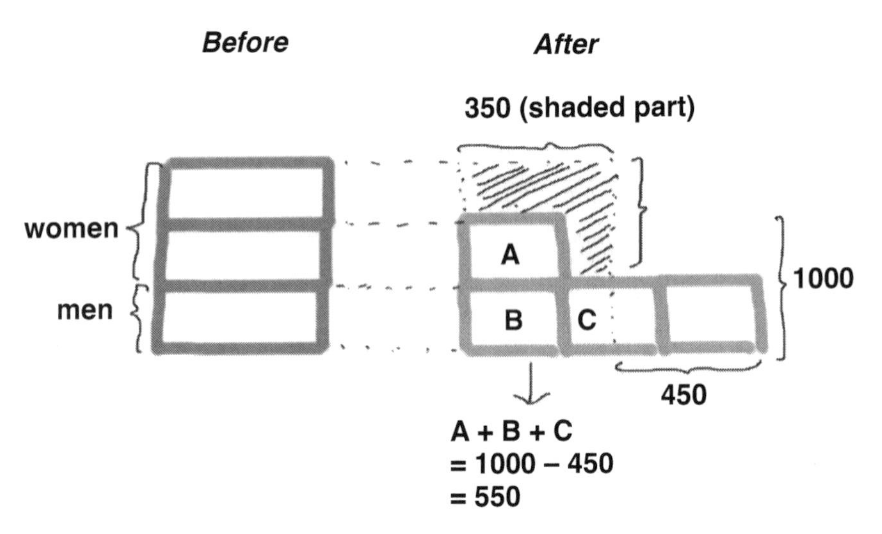

From the model,

number of adults at first = 350 + 550 = 900

Note: We needn't use the fact that there were twice as many women as men. *Why?* Since 350 women left and 450 men went in, there were an extra of 100 persons in the hall. So, there were 1000 – 100 = 900 adults at first.

Thought Process

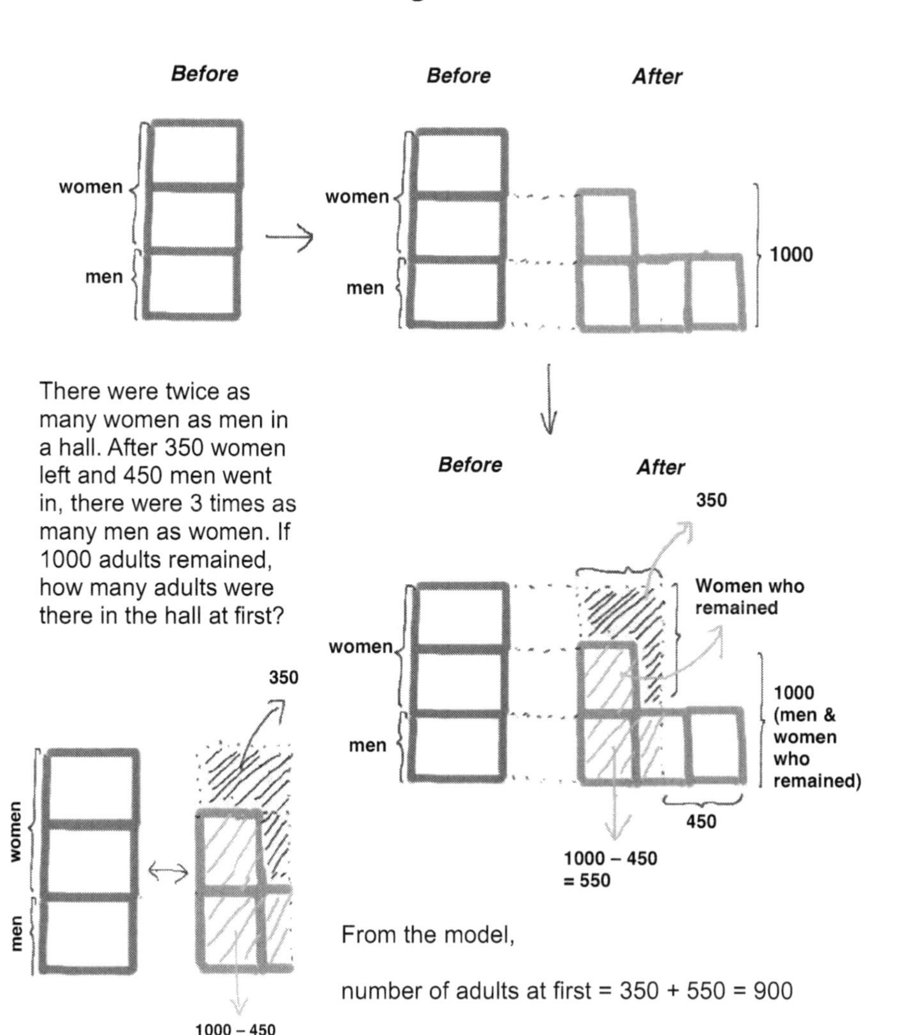

There were twice as many women as men in a hall. After 350 women left and 450 men went in, there were 3 times as many men as women. If 1000 adults remained, how many adults were there in the hall at first?

From the model,

number of adults at first = 350 + 550 = 900

Worked Example 29

There were twice as many women as men in a hall. After 350 women left and 450 men went in, there were 3 times as many men as women. If 1000 adults remained, how many adults were there in the hall at first?

Method 2

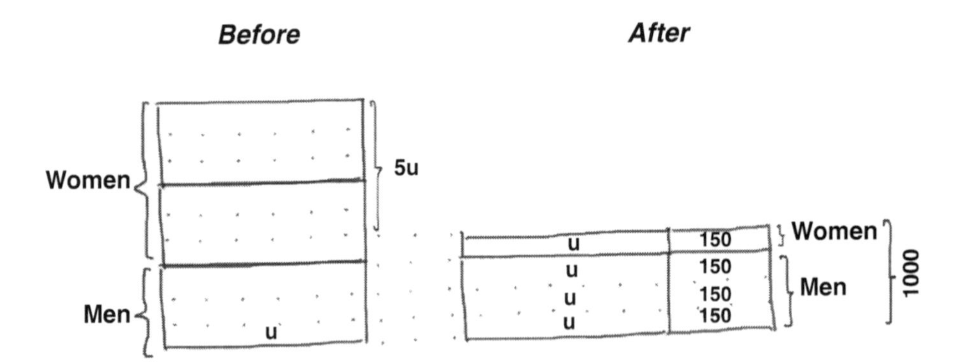

From the model,

4u + 4 × 150 → 1000
4u + 600 → 1000
4u → 1000 – 600 = 400
u → 400 ÷ 4 = 100
9u → 9 × 100 = 900

There are 900 adults in the hall at first.

Thought Process

There were twice as many women as men in a hall. After 350 women left and 450 men went in, there were 3 times as many men as women. If 1000 adults remained, how many adults were there in the hall at first?

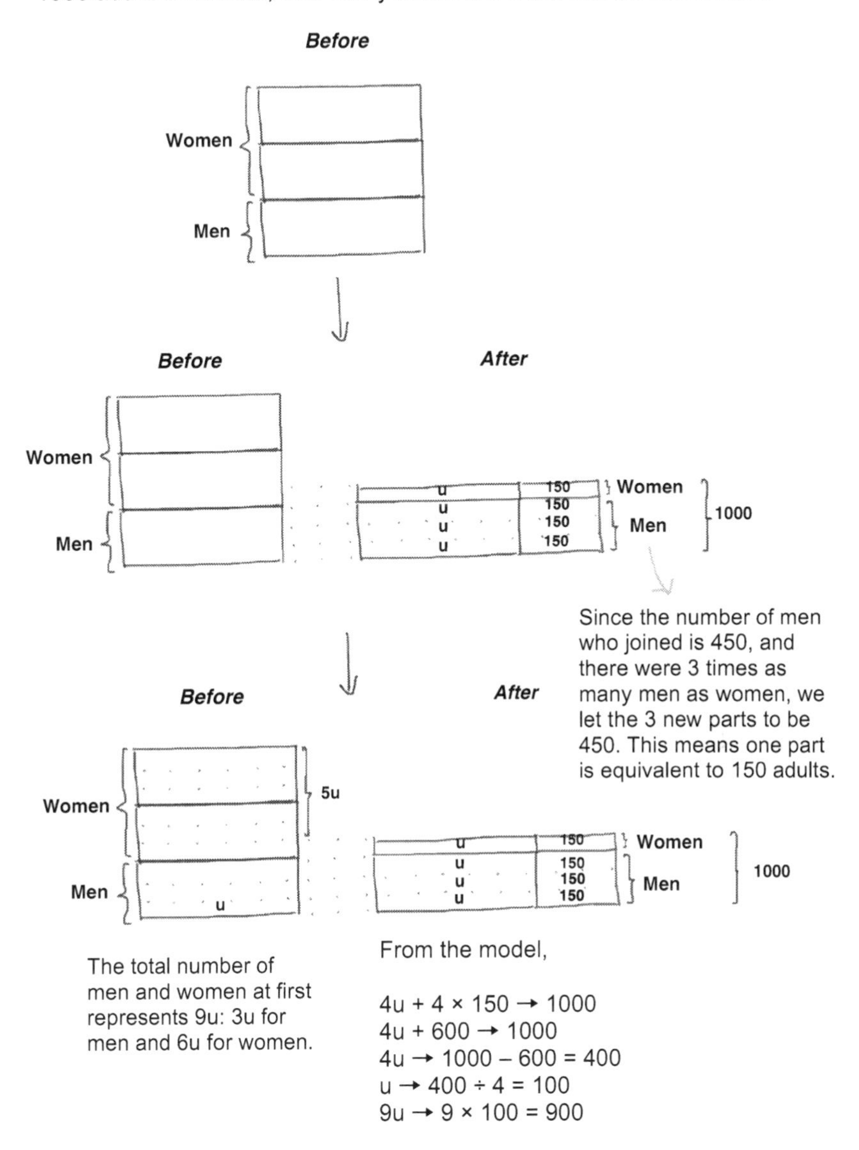

Since the number of men who joined is 450, and there were 3 times as many men as women, we let the 3 new parts to be 450. This means one part is equivalent to 150 adults.

The total number of men and women at first represents 9u: 3u for men and 6u for women.

From the model,

$4u + 4 \times 150 \rightarrow 1000$
$4u + 600 \rightarrow 1000$
$4u \rightarrow 1000 - 600 = 400$
$u \rightarrow 400 \div 4 = 100$
$9u \rightarrow 9 \times 100 = 900$

There are 900 adults in the hall at first.

Worked Example 29

There were twice as many women as men in a hall. After 350 women left and 450 men went in, there were 3 times as many men as women. If 1000 adults remained, how many adults were there in the hall at first?

Method 3

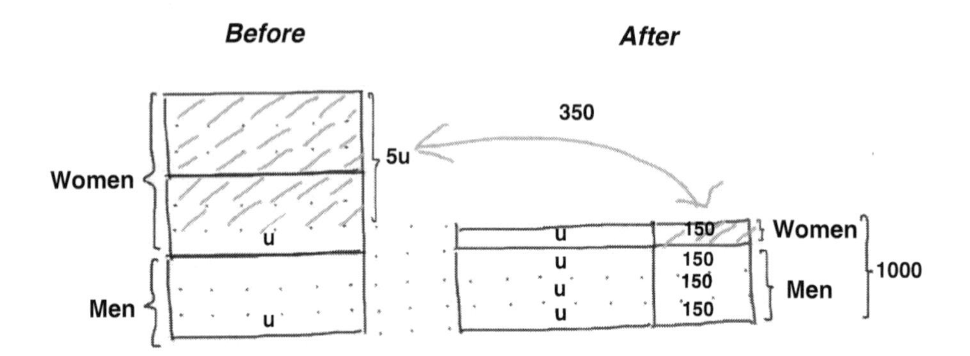

From the model,

5u → 350 + 150 = 500
u → 500 ÷ 5 = 100
9u → 9 × 100 = 900

There are 900 adults in the hall at first.

Worked Example 30

There were 3 times as many boys as girls in a museum. After 75 boys left and 25 girls entered, there were 2 times as many girls as boys. If 90 children remained, how many children were in the museum at first?

Method 1

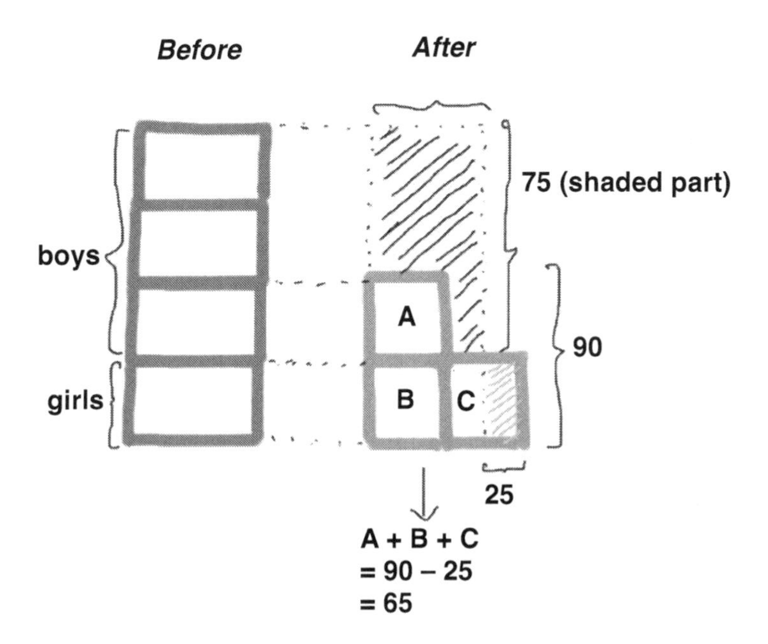

From the model,

number of children in the museum at first
= 75 + 65 = 140

Worked Example 30

There were 3 times as many boys as girls in a museum. After 75 boys left and 25 girls entered, there were 2 times as many girls as boys. If 90 children remained, how many children were in the museum at first?

Method 2

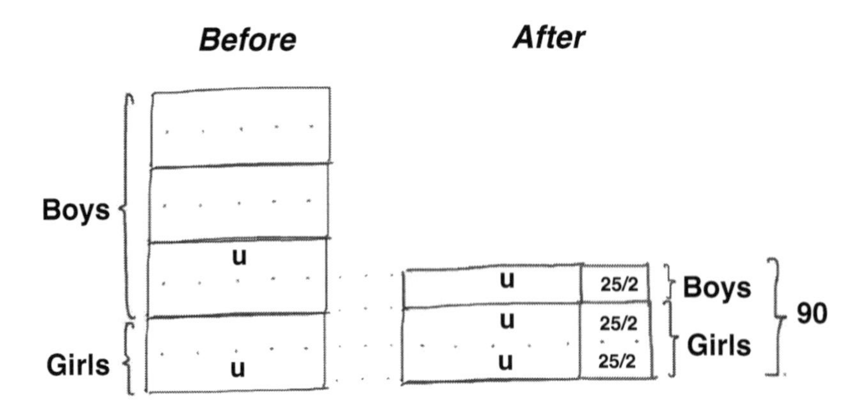

From the model,

$3u + 3 \times 25/2 \rightarrow 90$
$3u + 75/2 \rightarrow 90$
$3u \rightarrow 180/2 - 75/2 = 105/2$
$u \rightarrow 1/3 \times 105/2 = 35/2$
$8u \rightarrow 8 \times 35/2 = 140$

At first, 140 children were at the museum.

Why is the number 75, which is the number of boys who left, isn't used in the solution? Is it redundant?

Worked Example 30

There were 3 times as many boys as girls in a museum. After 75 boys left and 25 girls entered, there were 2 times as many girls as boys. If 90 children remained, how many children were in the museum at first?

Method 3

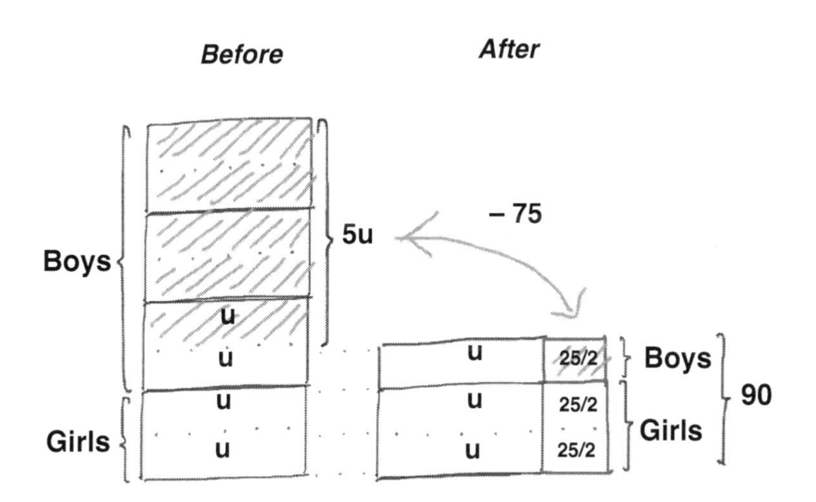

From the model,

$5u \rightarrow 75 + 25/2 = 150/2 + 25/2$
$= 175/2$
$u \rightarrow 1/5 \times 175/2 = 35/2$
$8u \rightarrow 8 \times 35/2 = 140$

At first, 140 children were at the museum.

Why is the number 90, which is the number of children who remained, isn't used in the solution? Is it redundant?

Worked Example 31

Nathan had half as many postcards as Albert. After Nathan gave 18 cards to Albert, Albert has 3 times as many postcards as Nathan. How many postcards did they have altogether?

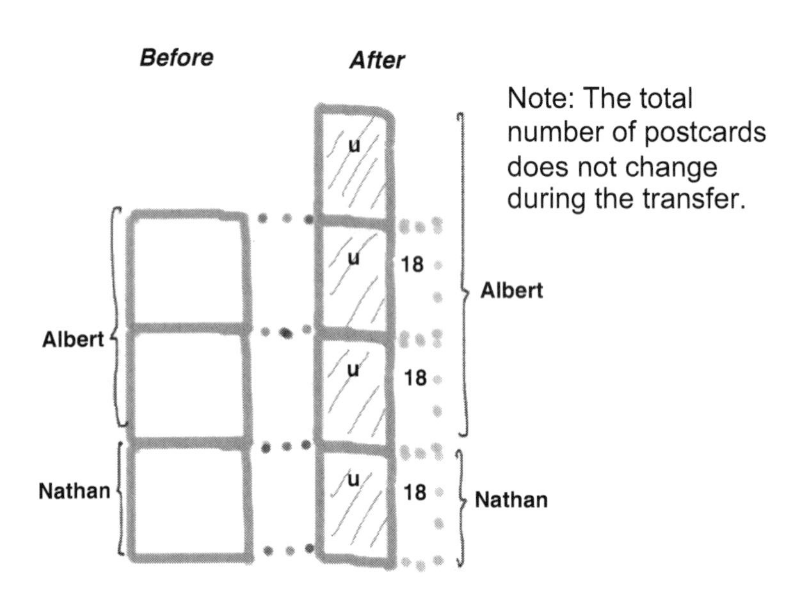

From the model,

u → 18 + 18 + 18 = 54
4u → 4 × 54 = 216

They had 216 postcards.

Check
Before:

Nathan: 54 + 18 = 72
Albert: 144 = 2 × 72

After:
Nathan: 54
Albert: 144 + 18 = 162 = 3 × 54

Compare this stack method with two other commonly used problem-solving heuristics: the bar method and the unitary method.

Which method do you prefer? Why?

Thought Process

Nathan had half as many postcards as Albert. After Nathan gave 18 cards to Albert, Albert has 3 times as many postcards as Nathan. How many postcards did they have altogether?

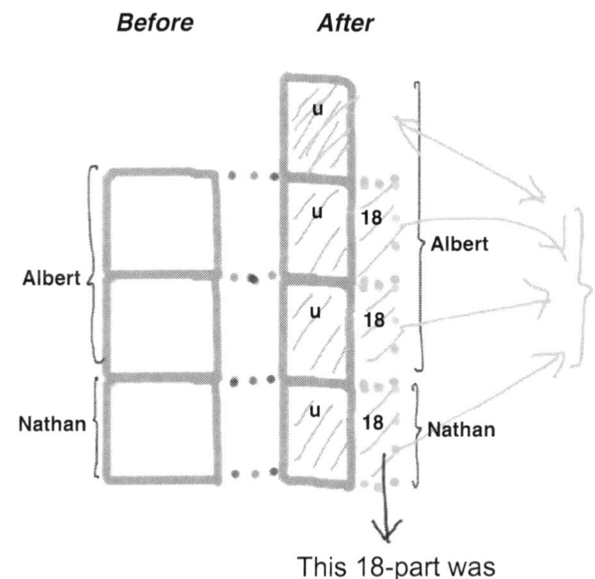

These three parts (18 + 18 + 18) were initially part of Nathan's and Albert's postcards.

Since the total number of postcards remains unchanged, this total of 54 postcards must belong to one (upper) of the three parts belonging to Albert after the transfer.

This 18-part was given to Albert.

From the model,

$u \rightarrow 18 + 18 + 18 = 54$
$4u \rightarrow 4 \times 54 = 216$

They had 216 postcards.

Worked Example 31

Nathan had half as many postcards as Albert. After Nathan gave 18 cards to Albert, Albert has 3 times as many postcards as Nathan. How many postcards did they have altogether?

Bar Model Method

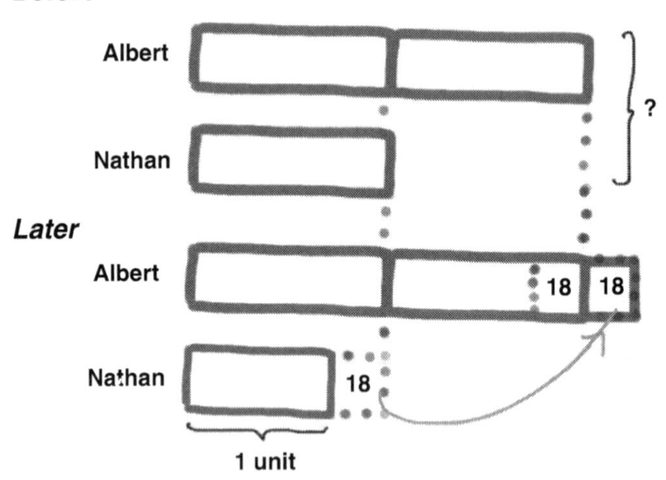

From the model,

1 unit → 18 + 18 + 18 = 54
3 units + 54 → 3 × 54 + 54 = 4 × 54 = 216

or
1 unit + 18 → 54 + 18 = 72
3 × 72 = 216

They had 216 postcards.

Worked Example 31

Nathan had half as many postcards as Albert. After Nathan gave 18 cards to Albert, Albert has 3 times as many postcards as Nathan. How many postcards did they have altogether?

Unitary Method

	Nathan	Albert	Total no. of cards
Before	1 unit	2 units	3 units
Change	– 18	+ 18	
After	1 part	3 parts	4 parts

	Nathan	Albert	Total no. of cards
Before	1 unit × 4 = 4 units	2 units × 4 = 8 units	12 units
Change	– 18	+ 18	
After	1 part × 3 = 3 units	3 parts × 3 = 9 units	12 units

From the table,

4 units – 3 units = 18
or
9 units – 8 units = 18
1 unit = 18
12 units = 12 × 18 = 216

They had 216 postcards.

Worked Example 32

Robin had 5 times as many stamps as Claire. After Claire bought 18 more stamps, Robin had now twice as many stamps as Claire. How many stamps did Robin have?

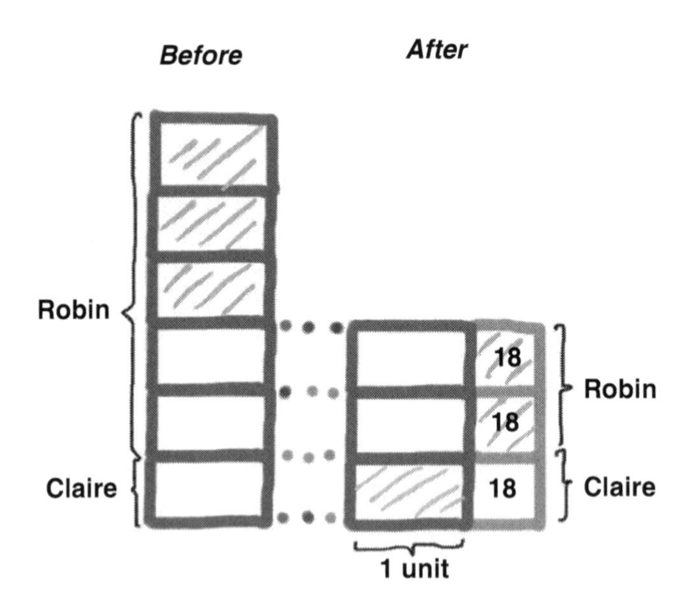

From the model,

3 units → 18 + 18 = 36
1 unit → 36 ÷ 3 = 12
5 units → 5 × 12 = 60

Robin had 60 stamps.

Thought Process

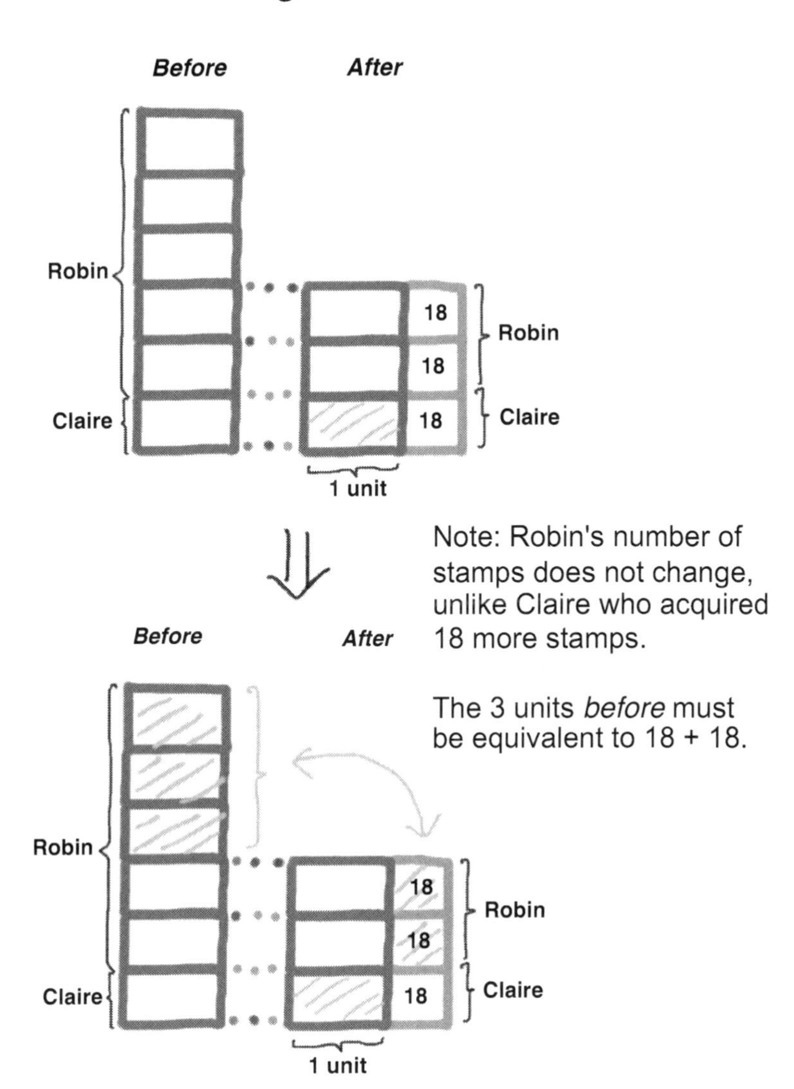

Note: Robin's number of stamps does not change, unlike Claire who acquired 18 more stamps.

The 3 units *before* must be equivalent to 18 + 18.

Worked Example 33

John had 3 times as many Singapore stamps as Malaysia stamps. After giving away 18 Malaysia stamps, he had 5 times as many Singapore stamps as Malaysia stamps. How many Singapore stamps did he have?

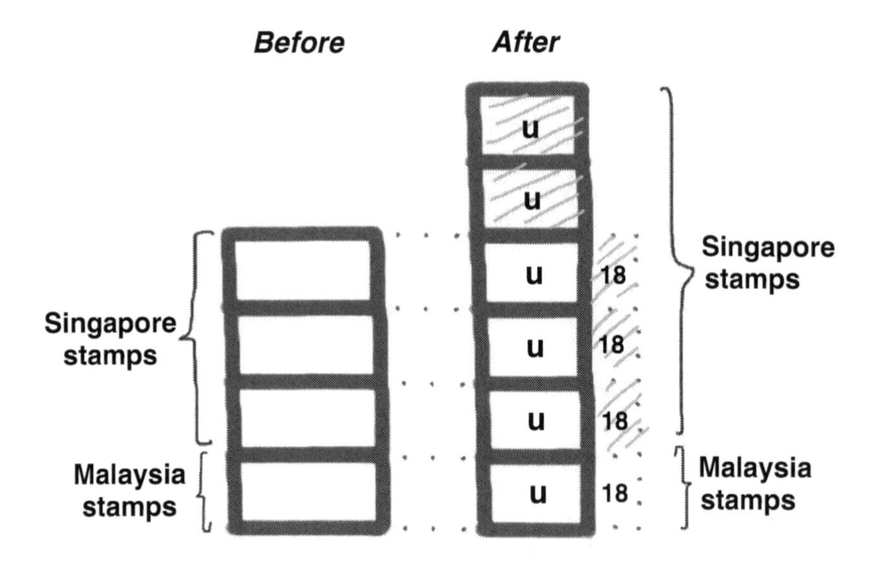

From the model,

2u → 18 + 18 + 18
u → 9 + 9 + 9 = 27
5u → 5 × 27 = 135

John had 135 Singapore stamps at first.

Note: The number of Singapore stamps remains unchanged.

Worked Example 34

Wilson and Ian had the same amount of money. After Wilson gave $60 to Ian, Ian had 4 times as much money as Wilson. How much money did they have altogether?

Method 1

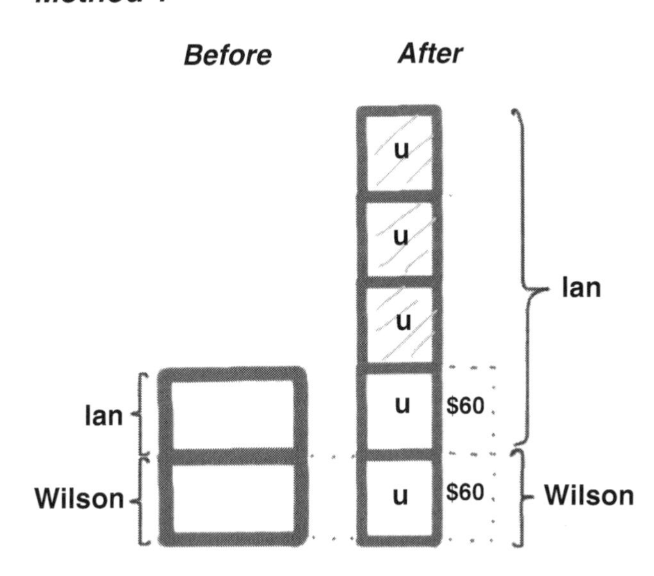

From the model,

3u → $60 + $60 = $120
u → $120 ÷ 3 = $40
5u → 5 × $40 = $200

Both had $200 altogether.

Note: The total amount of money remains unchanged during the transfer.

Worked Example 34

Wilson and Ian had the same amount of money. After Wilson gave $60 to Ian, Ian had 4 times as much money as Wilson. How much money did they have altogether?

Method 2

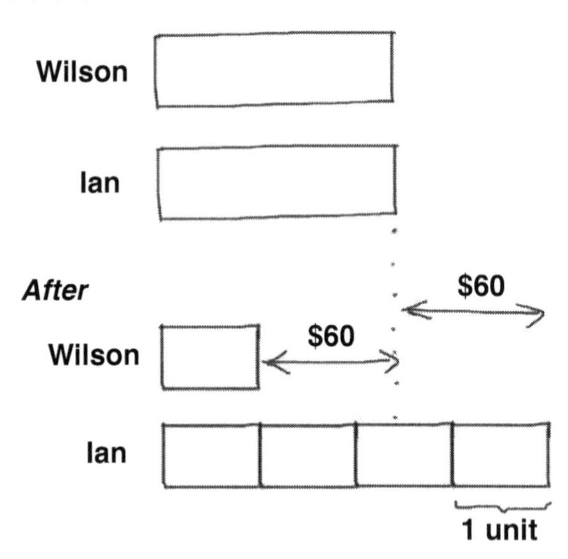

From the model,

3 units → $60 + $60 = $120
1 unit → $120 ÷ 3 = $40
5 units → 5 × $40 = $200

Ian and Wilson had $200 altogether.

Worked Example 34

Wilson and Ian had the same amount of money. After Wilson gave $60 to Ian, Ian had 4 times as much money as Wilson. How much money did they have altogether?

Method 3

	Wilson	Ian	Total
Before	? units	? units	5 units
Change	– $60	+ $60	
After	1 unit	4 units	5 units

From the table,
5 units ÷ 2 → $2\frac{1}{2}$ units
$2\frac{1}{2}$ units – 1 unit → 60
$1\frac{1}{2}$ units → 60
3/2 units → 60
1 unit → 60 × 2/3 = 40
5 units → 5 × 40 = 200

Note: The total amount of money remains unchanged during the transfer.

They had $200 in all.

Method 4 (By Logic)

Since both had the same amount of money, by giving $60 to Ian, Wilson will have 2 × $60 = $120 less than Ian.

Moreover, since Ian has 4 times as much money as Wilson, this means that Ian had 3 times more money than Wilson, which is equal to $120.

3 times → $120
1 time → $120 ÷ 3 = $40
5 times → 5 × $40 = $200

Both Wilson and Ian had $200 altogether.

Fractions

Fractions — Worked Examples 35–43

Examples involving fractions featured here are disguised versions of *before-after* or age-related word problems — the use of fractions instead of whole numbers.

These types of questions are often posed at higher levels under the topics of *Ratio* and *Percentage*. Again, these word problems lend themselves quite easily to the stack model method, which enable younger pupils to be exposed to higher-order mathematical thinking skills much earlier.

Worked Example 35

Pearl is 1/4 years the age of her mother. Pearl's mother is 1/2 the age of her grandfather. The sum of their ages is 104. How old is each person?

"Pearl is 1/4 years the age of her mother." implies that "Pearl's mother is 4 times as old as her".

Pearl's mother is 1/2 the age of her grandfather. This means that Pearl's grandfather is twice as old as her mother.

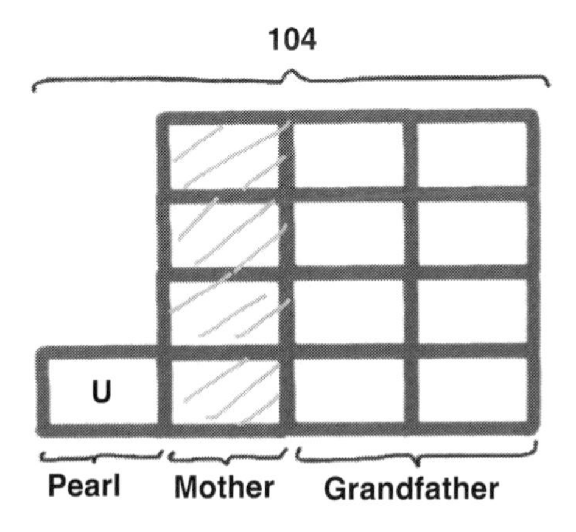

From the model,

13U → 104
U → 104 ÷ 13 = 8
4U → 4 × 8 = 32
8U → 8 × 8 = 64

Pearl is 8 years old.
Her mother is 32 years old.
Her grandfather is 64 years old.

Worked Example 36

A basket has some apples and 48 oranges. After 20 apples were sold, there were half as many apples as oranges. How many apples were there in the basket at first?

"... there were half as many apples as oranges" means that there were twice as many oranges as apples.

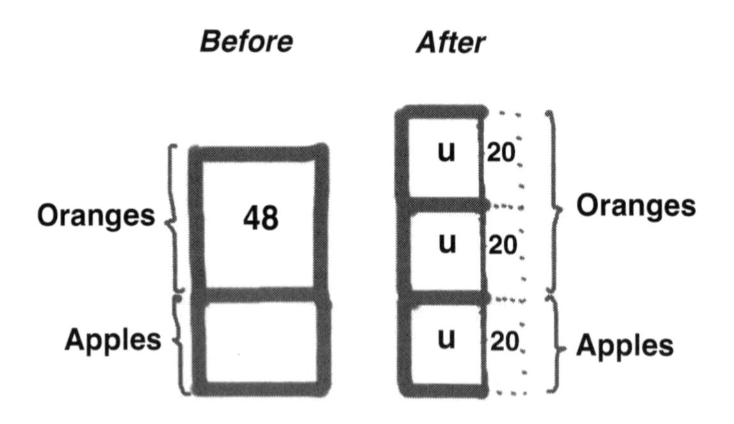

From the model,

Note: The number of oranges remains unchanged.

2u → 48
u → 48 ÷ 2 = 24
u + 20 → 24 + 20 = 44

The basket had 44 apples at first.

Worked Example 37

Eunice had 3/5 as many e-books as Warren. After Warren deleted 18 e-books, they both had the same number of e-books on their tablets. How many e-books did Warren have at first?

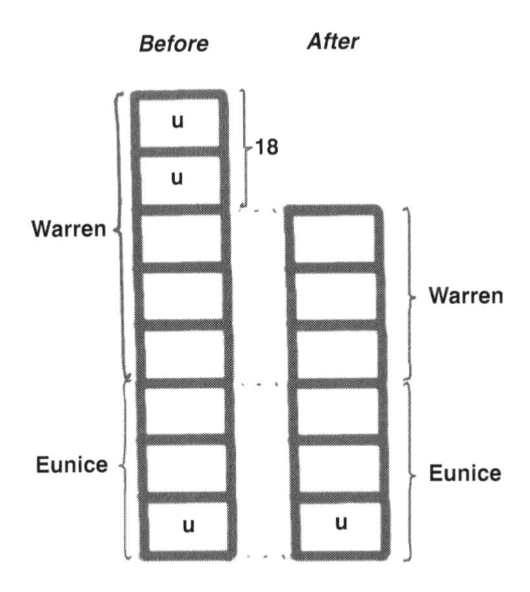

From the model,

2u → 18
u → 18 ÷ 2 = 9
5u → 5 × 9 = 45

Warren had 45 e-books at first.

Check

	Eunice	Warren
Before	3u → 27	45
After	27	45 – 18 = 27

Worked Example 38

When a father was 42, his son was 8. Now the son is one-third the father's age. How old is the son?

Method 1

"The son is 1/3 the father's age." implies that the father is 3 times as old as the son.

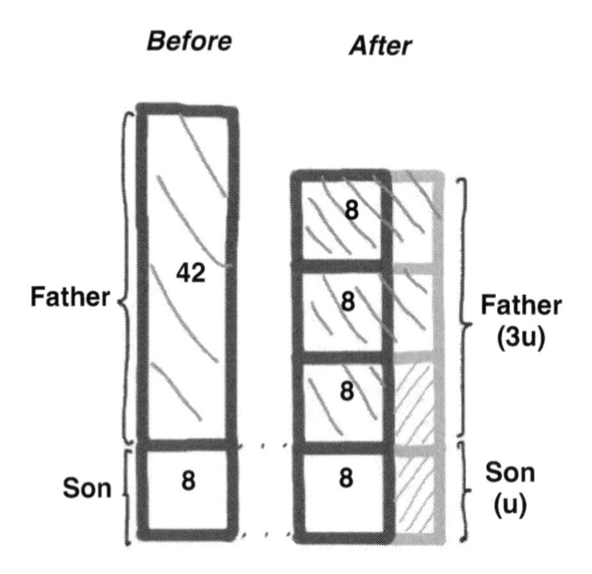

From the model,

$2u + 8 \rightarrow 42$
$2u \rightarrow 42 - 8 = 34$
$u \rightarrow 34 \div 2 = 17$

The son is 17 years now.

Worked Example 38

When a father was 42, his son was 8. Now the son is one-third the father's age. How old is the son?

Method 2

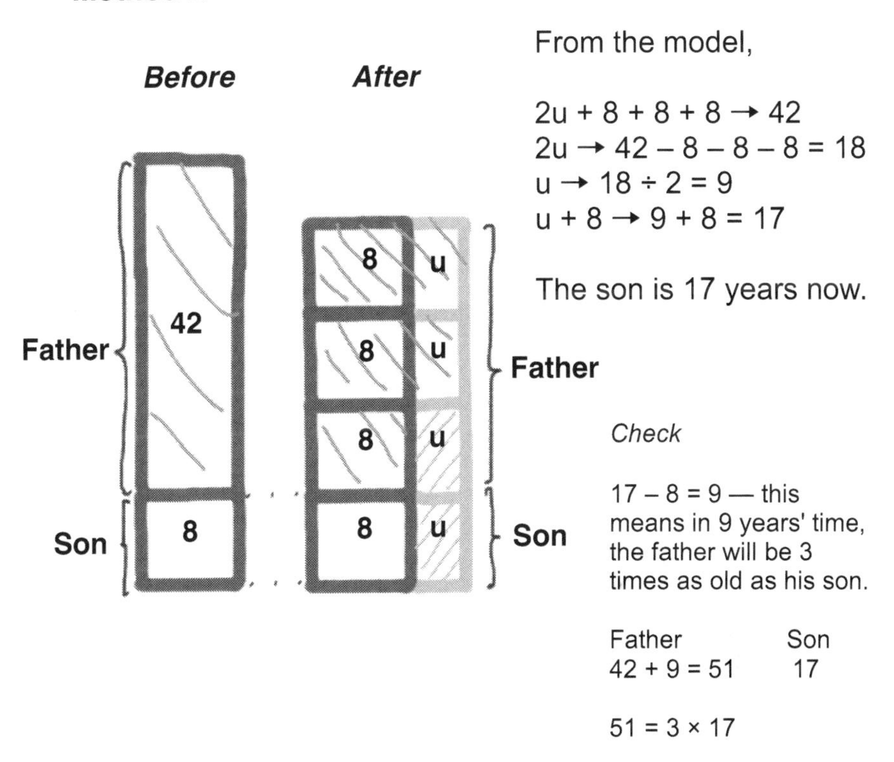

From the model,

$2u + 8 + 8 + 8 \rightarrow 42$
$2u \rightarrow 42 - 8 - 8 - 8 = 18$
$u \rightarrow 18 \div 2 = 9$
$u + 8 \rightarrow 9 + 8 = 17$

The son is 17 years now.

Check

$17 - 8 = 9$ — this means in 9 years' time, the father will be 3 times as old as his son.

Father	Son
$42 + 9 = 51$	17

$51 = 3 \times 17$

Alternatively, for both stack models in methods 1 and 2, you may also use the fact that the age difference between the father and the son does not change at any point in time, to solve the problem.

Worked Example 39

There is a piece of wire. One-third was used. After one-fifth of the remaining wire was used, 112 cm of the wire remained. What was the original length of the wire?

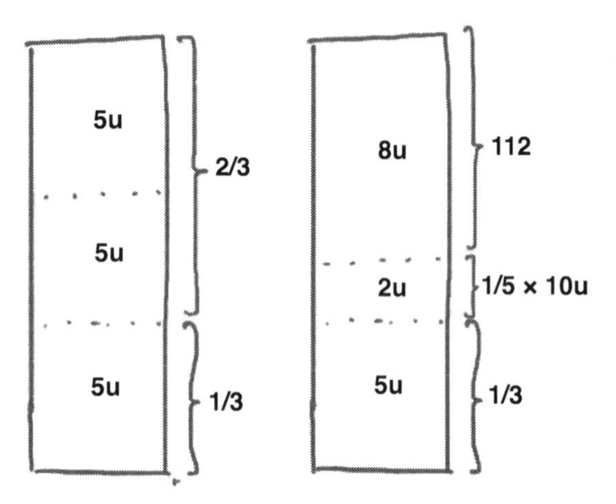

From the model,

8u → 112
u → 112 ÷ 8 = 14
15u → 15 × 14 = 30 × 7 = 210

The original length of the string was 210 cm.

Note that out of the 2/3 of the wire that remained, 1/5 of it was used.

From 2 out of 3 parts that were left, we need to take away 1/5 of it. So, if we divide the 2 parts into 2 × 5 units = 10 units, then 1/5 of the 10 units will represent 2 units. This means the remaining (10 – 2) = 8 units must represent 112 cm.

Worked Example 40

Mark has 1/8 as many marbles as John. After Mark receives 12 more marbles, he has half as many marbles as John. How many marbles does John have?

Method 1

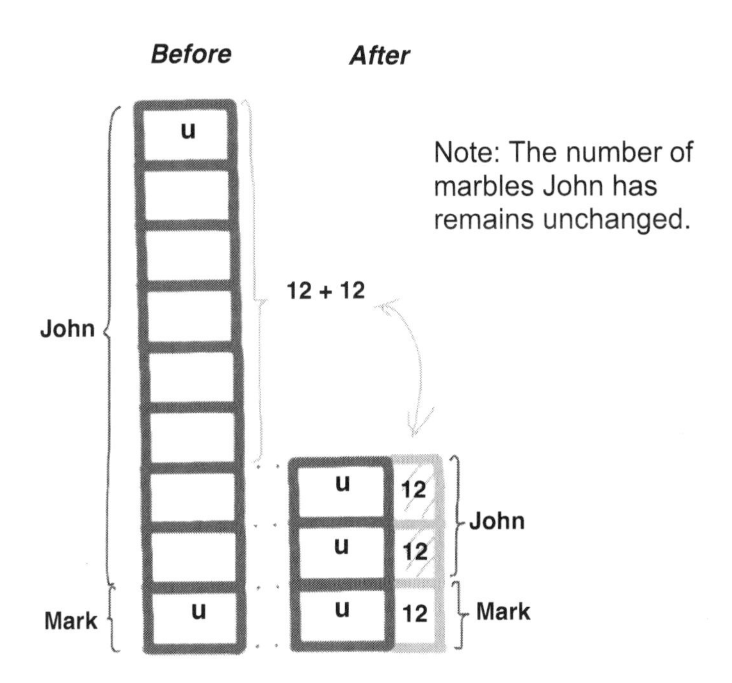

From the model,

$6u \rightarrow 12 + 12 = 24$
$u \rightarrow 24 \div 6 = 4$
$8u \rightarrow 8 \times 4 = 32$

John has 32 marbles.

Worked Example 40

Mark has 1/8 as many marbles as John. After Mark receives 12 more marbles, he has half as many marbles as John. How many marbles does John have?

Method 2

	Mark	**John**
Before	1 unit	8 units
Change	+ 12	
After	1 part × 4 = 4 units	2 parts × 4 = 8 units

From the table,

4 units – 1 unit → 12
3 units → 12
1 unit → 12 ÷ 3 = 4
8 units → 8 × 4 = 32

John has 32 marbles.

Worked Example 41

Farmer Yan had twice as many cows as sheep. After he had sold 369 cows and another 24 cows died, he had half as many cows as sheep left. How many cows remained?

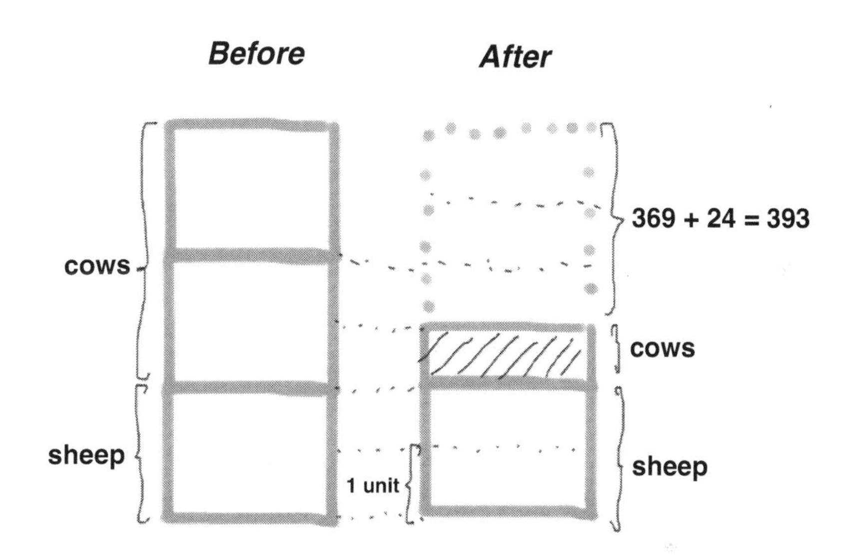

From the model,

3 units → 369 + 24 = 393
1 unit → 393 ÷ 3 = 131

131 cows remained.

Check:
Sheep before or after = 2 × 131 = 262

Cows before = 2 × 262 = 524
Cows after = 524 – 393 = 131
= 1/2 × 262

Worked Example 42

The total number of Facebook friends and fans is 156. Three-fifths of the Facebook fans belong to Joan, and 1/3 of the Facebook friends belong to Richard. Richard has 8 more friends than fans. How many Facebook friends does Joan have?

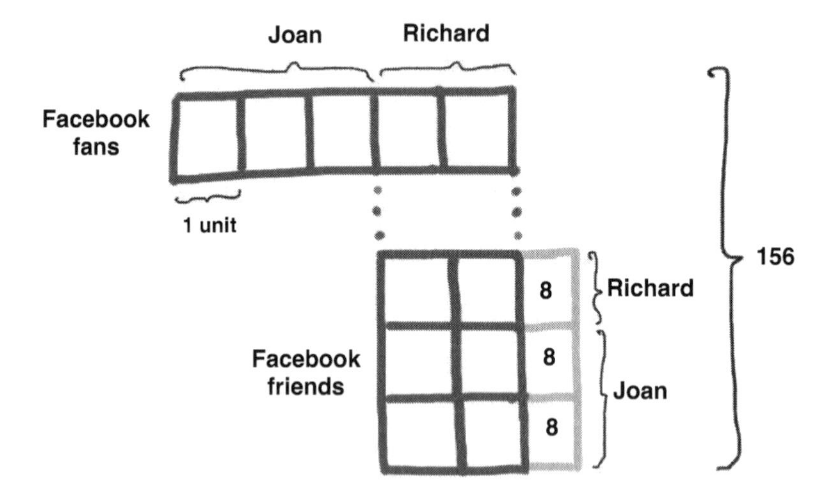

From the model,

11 units + 8 + 8 + 8 → 156
11 units → 156 – 24 = 132
1 unit → 132 ÷ 11 = 12
4 units + 8 + 8 → 4 × 12 + 16 = 48 + 16 = 64

Joan has 64 Facebook friends.

Worked Example 43

At a birthday party, the number of girls was 3/5 the number of boys. After 18 girls left the room, the number of girls remaining was 3/10 the number of boys. How many boys were at the party?

Method 1

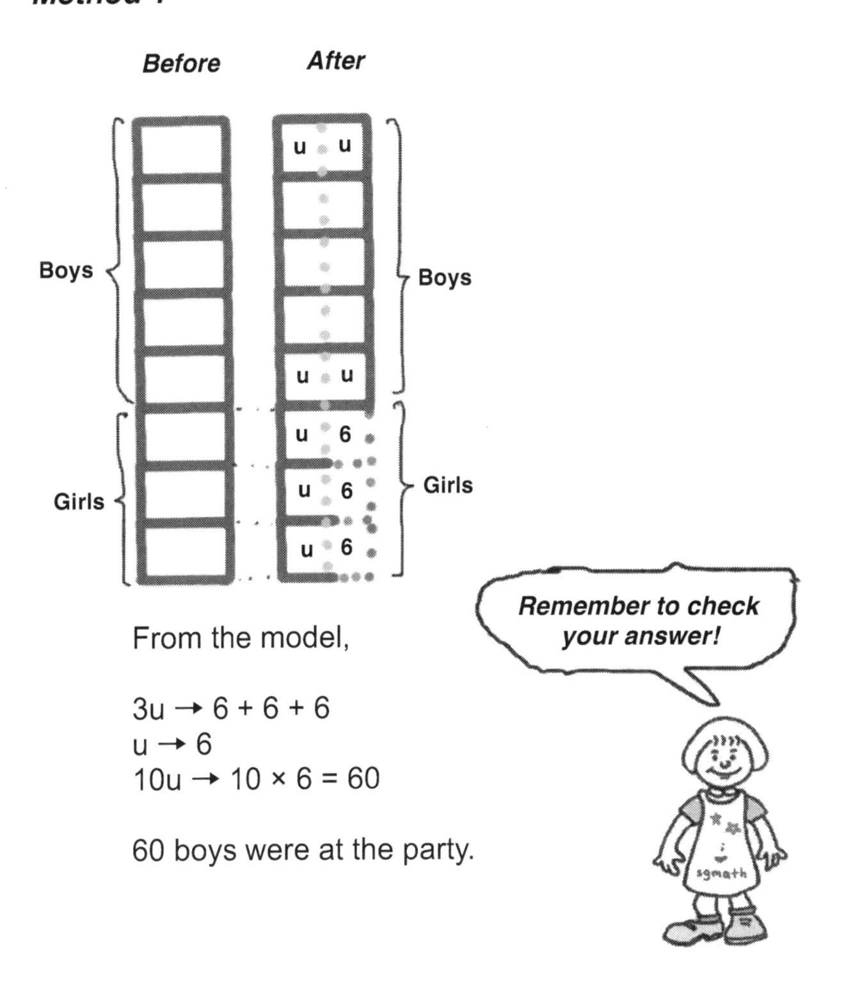

From the model,

3u → 6 + 6 + 6
u → 6
10u → 10 × 6 = 60

60 boys were at the party.

Thought Process

At a birthday party, the number of girls was 3/5 the number of boys. After 18 girls left the room, the number of girls remaining was 3/10 the number of boys. How many boys were at the party?

Method 1

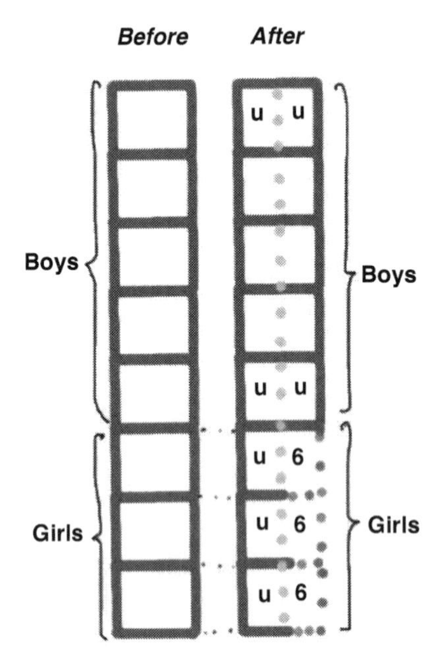

Observe that since there is no change in the number of boys, the 5 parts belonging to the boys before is now equivalent to the 10 units after 18 girls left. In other words, 1 part is equivalent to 2 units.

After 18 girls left, the girls now represent 3 units, while the boys still represent 10 units.

At first, the 3 parts that define the girls represent $3 \times 2 = 6$ units, of which only 3 units remain. So, the other 3 units must represent the 18 girls who left, or 1 unit represents 6 girls.

Since the boys represent 5 parts or 10 units, there were $10 \times 6 = 60$ boys at the party.

Worked Example 43

At a birthday party, the number of girls was 3/5 the number of boys. After 18 girls left the room, the number of girls remaining was 3/10 the number of boys. How many boys were at the party?

Method 2

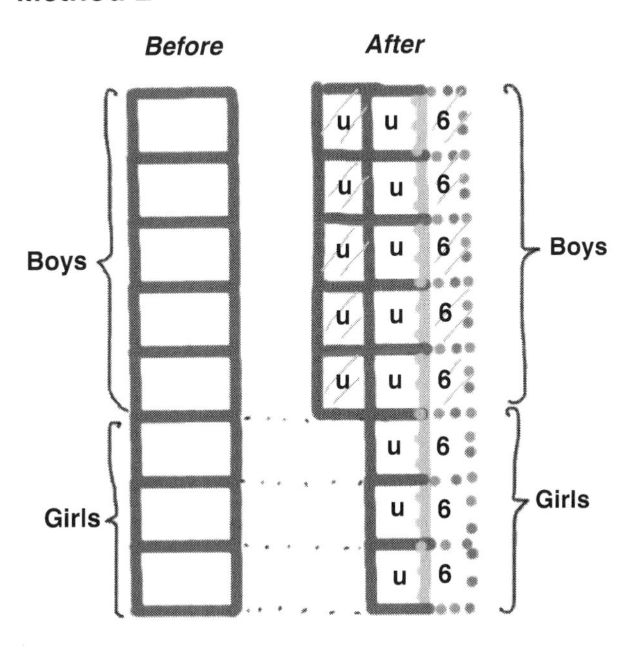

From the model,

5u → 5 × 6
u → 6
5u + 5 × 6 → 30 + 30 = 60
or 10u → 10 × 6 = 60

60 boys were at the party.

Observe that the total number of boys remains unchanged.

Worked Example 43

At a birthday party, the number of girls was 3/5 the number of boys. After 18 girls left the room, the number of girls remaining was 3/10 the number of boys. How many boys were at the party?

Method 3

	Girls	Boys
Before	3 units	5 units
Change	– 18 girls	
After	3 parts	10 parts

Since the number of boys remains unchanged, 5 units = 10 parts.

↓

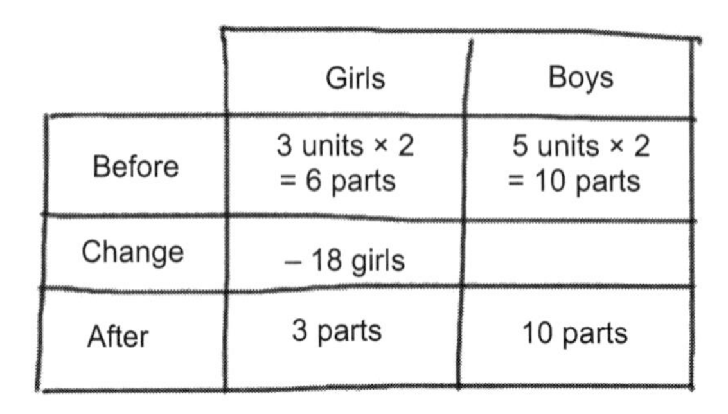

	Girls	Boys
Before	3 units × 2 = 6 parts	5 units × 2 = 10 parts
Change	– 18 girls	
After	3 parts	10 parts

From the table,

6 parts – 3 parts → 18
3 parts → 18
1 part → 18 ÷ 3 = 6
10 parts → 10 × 6 = 60

60 boys were at the party.

Decimals & Money

Decimals and Money — Worked Examples 44–50

The examples here show that both routine and non-routine questions involving decimals may be solved quite easily using the stack model method.

Again, these questions use similar reasoning or problem-solving skills that were previously applied to solve word problems on *Whole Numbers* and *Fractions*.

Compare the given stack-model solutions with other problem-solving heuristics. *Which one do you prefer?*

Worked Example 44

Joel and Gina have $62.80 in total. Joel has $18.80 more than Gina. How much money does each of them have?

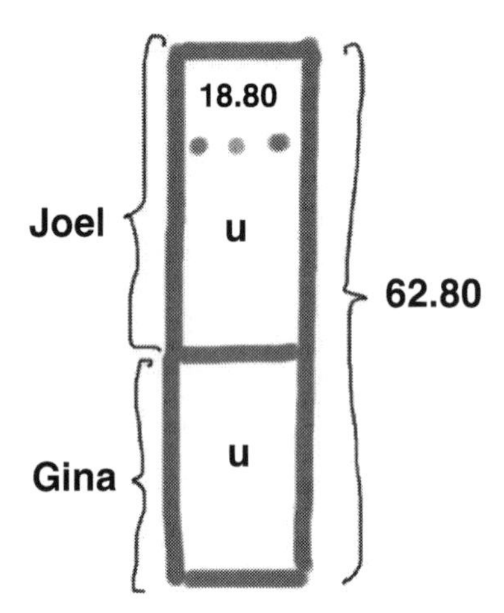

From the model,

2u → 62.80 – 18.80 = 44
u → 44 ÷ 2 = 22

u + 18.80 = 22 + 18.80 = 40.80

Gina has $22 and Joel has $40.80.

Practice

In total, Jane and Walter have $58.10. Walter has $9.90 more than Jane. How much does each of them have?

Answer: Jane: $24.10; Walker: $34.

Worked Example 45

The total savings of Jenny and Irene are $51.35. If Jenny saves $23.80, how much more money does Irene save than Jenny?

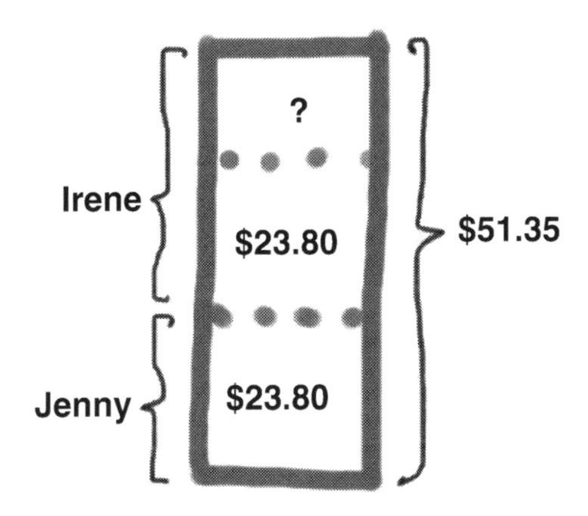

From the model,

23.80 + 23.80 = 47.60
51.35 – 47.60 = 3.75

Irene saves $3.75 more than Jenny.

Observe that we needn't find the amount of money Irene saves first in order to find the answer.

Note: Although the topic on decimals is not formally taught in primary 3 and 4, however, it is not uncommon for P4 pupils to learn it in an informal setting.

Worked Example 46

Beth and Dave had $220 altogether. After each donated $50 to charity, Beth had 3 times as much money left as Dave had left. How much did Beth have at first?

Method 1

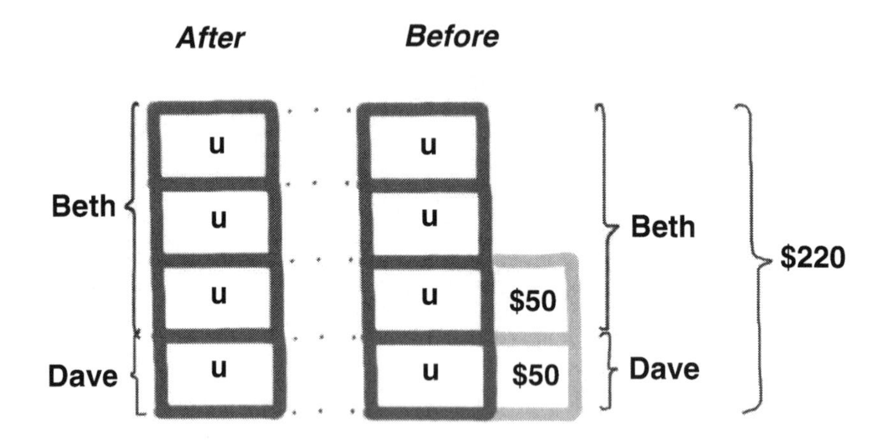

From the model,

4u + $50 + $50 → $220
4u → $220 – $50 – $50 = $120
u → $120 ÷ 4 = $30
3u + $50 → 3 × $30 + $50 = $90 + $50 = $140

Beth had $140 at first.

Worked Example 46

Beth and Dave had $220 altogether. After each donated $50 to charity, Beth had 3 times as much money left as Dave had left. How much did Beth have at first?

Method 2

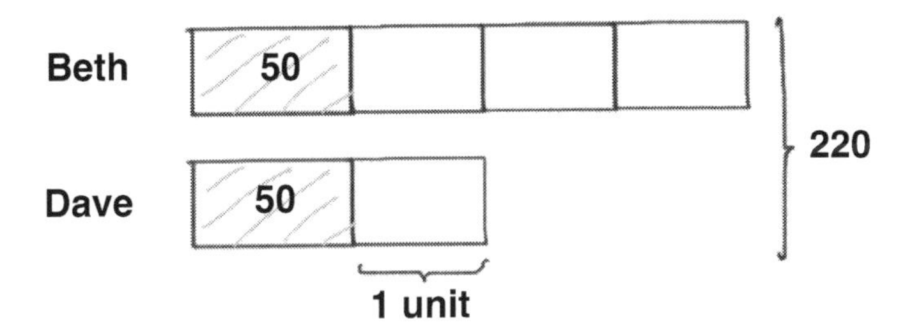

From the model,

4 units → 220 – 50 – 50 = 120
1 unit → 120 ÷ 4 = 30
3 units + 50 → 3 × 30 + 50 = 90 + 50 = 140

Beth had $140 at first.

Worked Example 46

Beth and Dave had \$220 altogether. After each donated \$50 to charity, Beth had 3 times as much money left as Dave had left. How much did Beth have at first?

Method 3

Beth	**Dave**
\$220	
– \$50	– \$50
3 :	1

Beth	**Dave**
△3	△1
+ \$50	+ \$50
\$220	

△3 + △1 = \$220 – \$50 – \$50 = \$120

4 △1 = \$120

△1 = \$120 ÷ 4 = \$30

△3 + \$50 = 3 × \$30 + \$50 = \$140

Beth had \$140 at first.

Worked Example 47

Roy had 3 times as much money as Jane. After Roy gave $165 to Jane, he had twice as much money as she did. How much money did Roy have at first?

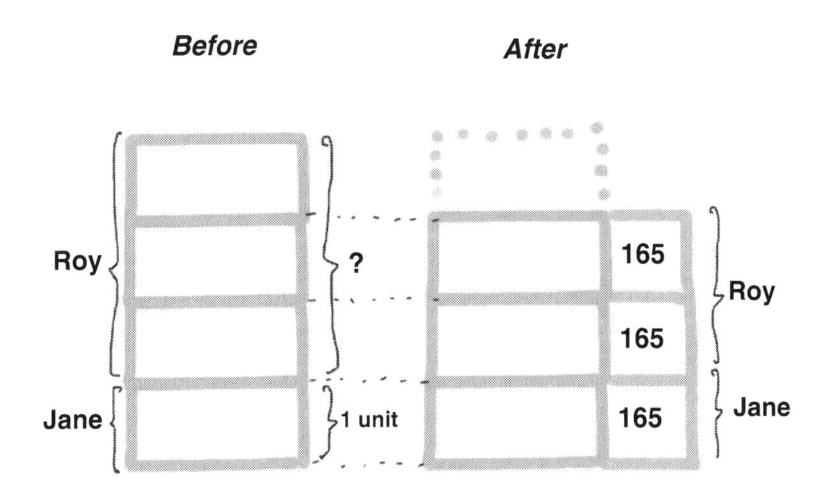

From the model,

1 unit → 3 × 165
3 units → 3 × 3 × 165 = 1485

Roy had $1485 at first.

Worked Example 48

Roy and Sally each had the same amount of money at first. After Roy earned $116 from ads to his blog, and Sally spent $20 to buy a WordPress template, Roy had now five times as much money as Sally. How much money did Sally have at first?

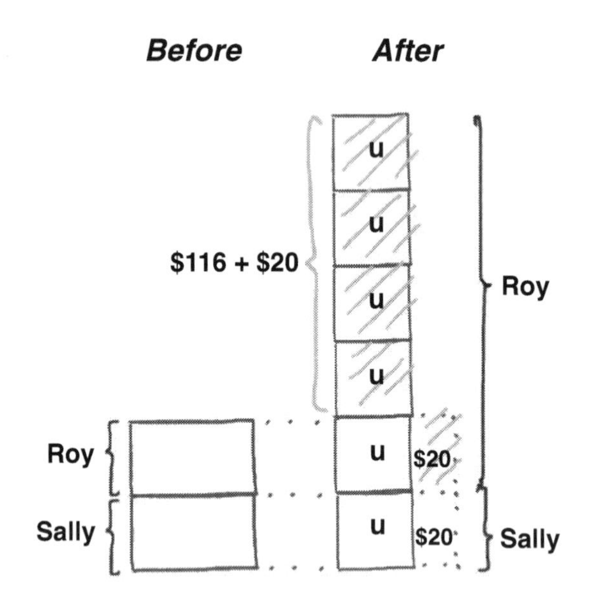

From the model,

4u → $116 + $20 = $136
u → $136 ÷ 4 = $34
u + $20 → $34 + $20 = $54

Sally had $54 at first.

Worked Example 49

Ian has twice as many coins as Kate. But if Ian were to give away 20 coins, Kate would have three times as many as Ian, how many coins does Ian have?

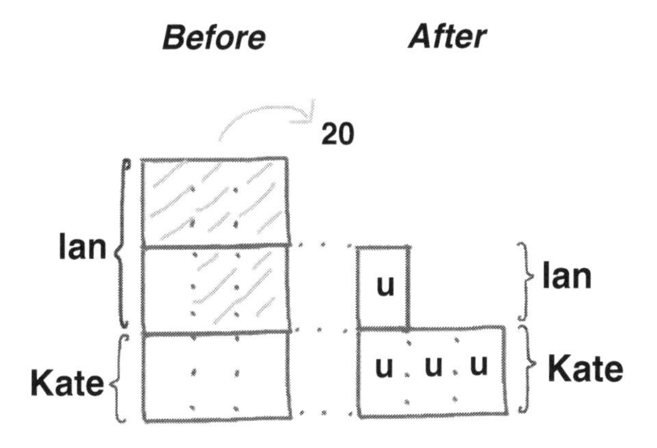

From the model,

5u → 20
u → 20 ÷ 5 = 4
6u → 6 × 4 = 24

Ian has 24 coins.

Check

	Before	After
Ian	6u → 24	24 – 20 = 4
Kate	3u → 12	12 = 3 × 4

Worked Example 50

Ian has twice as many coins as Kate. If Ian were to give Kate 20 coins, Kate would have three times as many as Ian. How many coins does Ian have?

Method 1

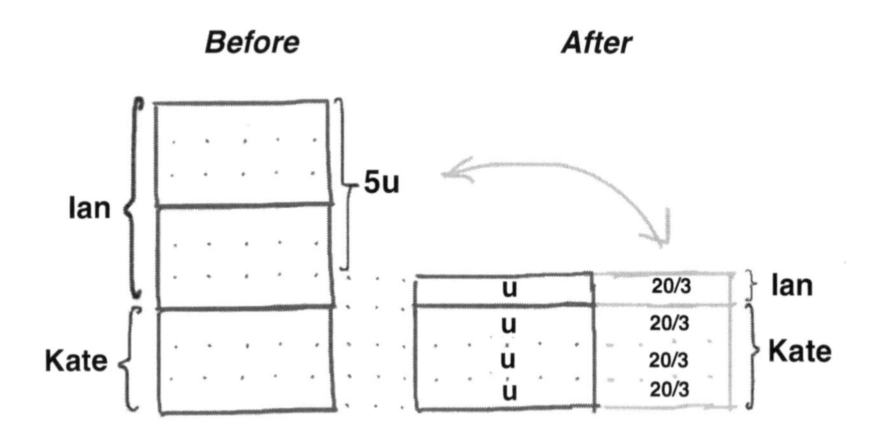

From the model,

5u → 4 × 20/3
u → 1/5 × 4 × 20/3 = 16/3
6u → 6 × 16/3 = 32

Ian has 32 coins.

Check
Ian: 32 – 20 = 12
Kate: 16 + 20 = 36 = 3 × 12

Kate would have 3 times as many coins as Ian.
Ian → 1 part
Kate → 3 parts

3 parts → 3u + 20
1 part → u + 20/3

Worked Example 50

Ian has twice as many coins as Kate. If Ian were to give Kate 20 coins, Kate would have three times as many as Ian. How many coins does Ian have?

Method 2

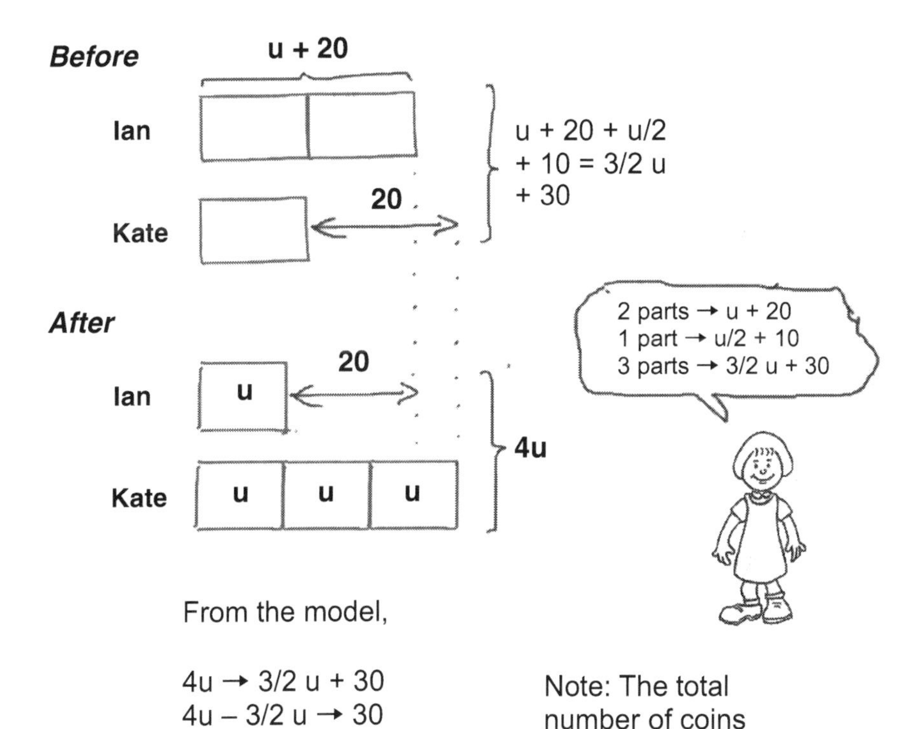

From the model,

$4u \rightarrow 3/2\ u + 30$
$4u - 3/2\ u \rightarrow 30$
$5/2\ u \rightarrow 30$
$u \rightarrow 30 \times 2/5 = 12$
$u + 20 \rightarrow 12 + 20 = 32$

Ian has 32 coins.

Note: The total number of coins *before* and *after* remains unchanged.

Worked Example 50

Ian has twice as many coins as Kate. If Ian were to give Kate 20 coins, Kate would have three times as many as Ian. How many coins does Ian have?

Method 3 (Optional)

Without the bar- or stack-model method, you would probably have to wait until secondary one to be able to solve this question, using algebra.

An algebraic approach would be as follows:

Ian	Kate
$2x$	x
$2x - 20$	$x + 20$

Given:

$$x + 20 = 3(2x - 20)$$
$$x + 20 = 6x - 60$$
$$20 + 60 = 6x - x$$
$$80 = 5x$$
$$x = 80/5 = 160/10 = 16$$
$$2x = 2 \times 16 = 32$$

Ian has 32 coins.

Revision Questions

Here is a sample of **25 miscellaneous primary 3–4 word problems** that may be solved using the stack model method.

The aim here is to encourage the pupils to solve the questions the "Stack Model Way", then to compare their solutions against other problem-solving heuristics, such as the bar model and Sakamoto methods. *Which types of word problems are favourable to the stack model method?*

1. Farmer John has 720 sheep. He shears half of them on Monday and two-thirds of the remainder on Tuesday. How many are left to be shared on Wednesday?

2. One whole brick weighs 5 kg and half a brick. How much does each brick weigh?

3. A rope 50 metres long is cut into two pieces.
   One piece is 3/7 the length of the other.
   How long is the shorter piece?

4. *Singapura Mathemetica* has 212 pages. The day I bought it I read exactly half. The next day I read one-third of what was left. On the day after that I read one-quarter of what remained. What fraction of the Singapore maths playbook remained unread?

Mathedoddles, anyone?

Addition

Degrees

Parallel

Hint: Do you need to find the number of pages read every day?

5. A box with 5 identical balls weighs 10.0 kg. The same box with 8 identical balls weighs 14.5 kg. How much does the box weigh?

If x is not the mass of the box, then …

6. Mark bought some milk bars. He ate 5 milk bars and divided the rest into 3 equal shares. She gave 1 share and 6 more milk bars to Mary. She was left with 14 milk bars. How many milk bars did Mark buy at first?

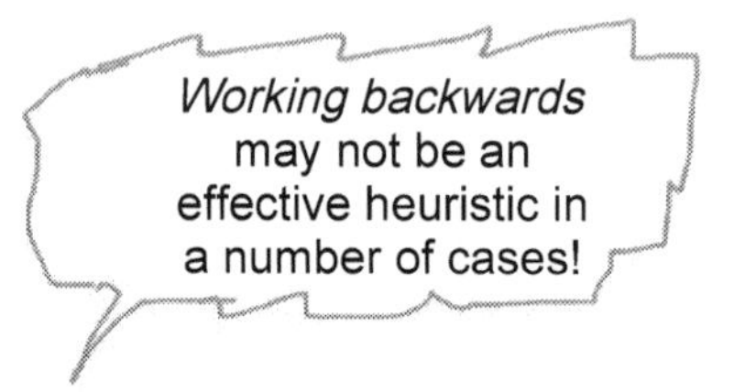

7. At a maths talk, there were only single mothers and married women with their husbands. If 1/3 of the women at the talk were single mothers, what fraction of the people present were married men?

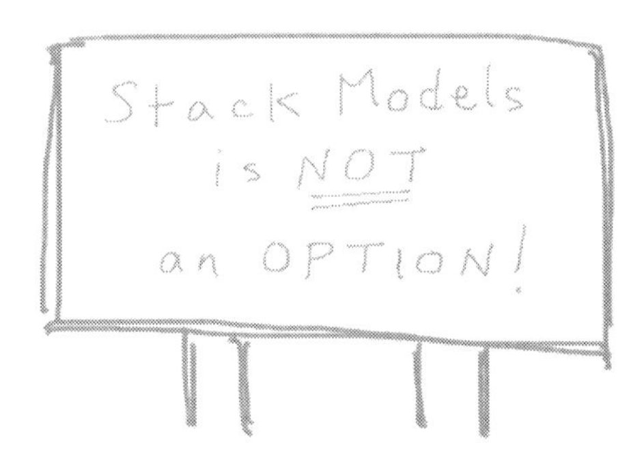

Hint: The number of husbands must be equal to the number of married women.

8. Carl poured water into two containers A and B. At first, the volume of water in container A was 4/5 the volume of water in container B. After 180 cm^3 of water from container A was poured into container B, the volume of water in container A was 1/2 the volume of water in container B. How much water was in container A at first?

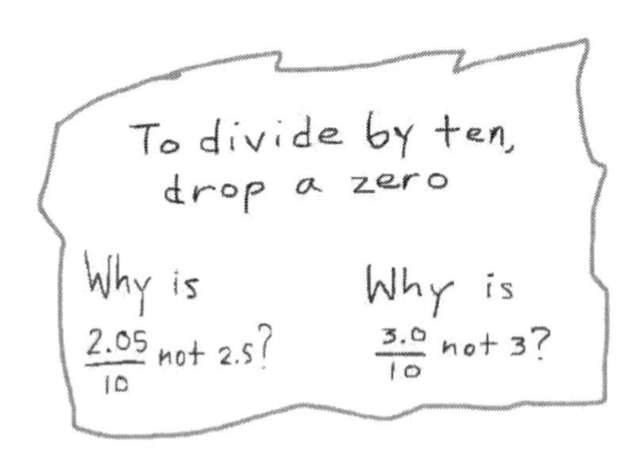

Hint: The total volume of water in both containers remains unchanged during the transfer.

*9. Doris, Esther and Fiona have 55 Facebook fans. They each have more than 10 Facebook fans. Doris has twice as many fans as Esther, and Fiona has the least number of fans. How many Facebook fans does Esther have?

Mathematical
Graphiti

BAR modeling

the mathematics
of parts and wholes

Hint: The total number of fans Doris and Esther have must be a multiple of 3 — (2 parts for Doris and 1 part for Esther).

10. Leo has 2/3 as many books as Jill. If Jill gives Leo 8 books, they both will have the same number of books. How many books do both persons have altogether?

Singapore MATH is not...

A Problem-Solving Heuristic ☑

A Product or Service ☑

A Trademark ☑

A Syllabus ☑

A Maths Curriculum ☑

Bar Modelling ☑

A Set of Books ☑

Hint: The total number of books does not change during the transfer.

11. One-third of the people at a meeting are women, a quarter are girls, one-sixth are men and there are six boys. How many people were at the party?

WHAT IF MY MATHS TEACHER IS A BIT WEIRD?

12. Oliver saved $70 more than Sally. Karen saved 6 times as much money as Sally. Their total savings is $830. How much money did Karen save?

The Stack Model Method

Look-See Proofs 4 Kids

11. Hint: If the number of people at the party represents one whole, how many parts must you divide the whole, so that the number of parts can be divided by 3, 4 and 6, without any remainders?

13. Henry has twice as much money as Mary in their PayPal accounts. After Henry used the PayPal app to buy an e-book costing $30, Mary has now three times as much money as Henry in their accounts. How much money did Henry have in his PayPal account at first?

Solve this problem, using both the Stack Model and the Unitary methods. Which one is easier? Why?

14. Rashid is 10 years younger than Alison. Louis is 3 times as old as Rashid. Their combined ages is 90 years. How old was Alison 12 years ago?

Singapura Sudoku

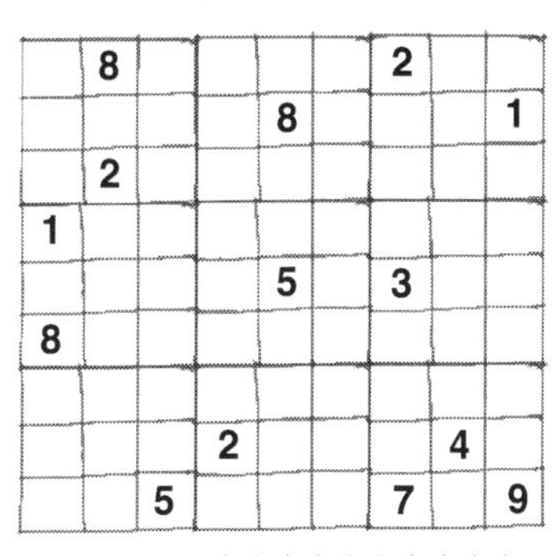

Difficulty Rating: ☆☆☆☆☆☆☆☆☆☆

15. At present, Joelle is 5 years old, and her mother is 45 years old. How many years from now will Joelle's mother be six times as old as her?

Hint: See Worked Examples 17 and 18.

16. The sum of the ages of Lisa and Ben is 76. Ben is three times as old as Lisa. In how many years' time will Ben be twice as old as Lisa?

Solve this problem, using at least two problem-solving heuristics.

Hint: How old are Lisa and Ben now? See Worked Examples 17–19, depending on which method you wish to use to solve this problem.

17. Samuel is 2 times and an additional 6 years as old as Moses. 12 years ago, Samuel was 4 times as old as Moses. How old is Samuel now?

18. Alice and Mike downloaded the same number of photos on Instagram. After Alice deleted 32 photos and Mike deleted 14 photos, Mike had four times as many photos left as Alice. How many photos did each person download in the beginning?

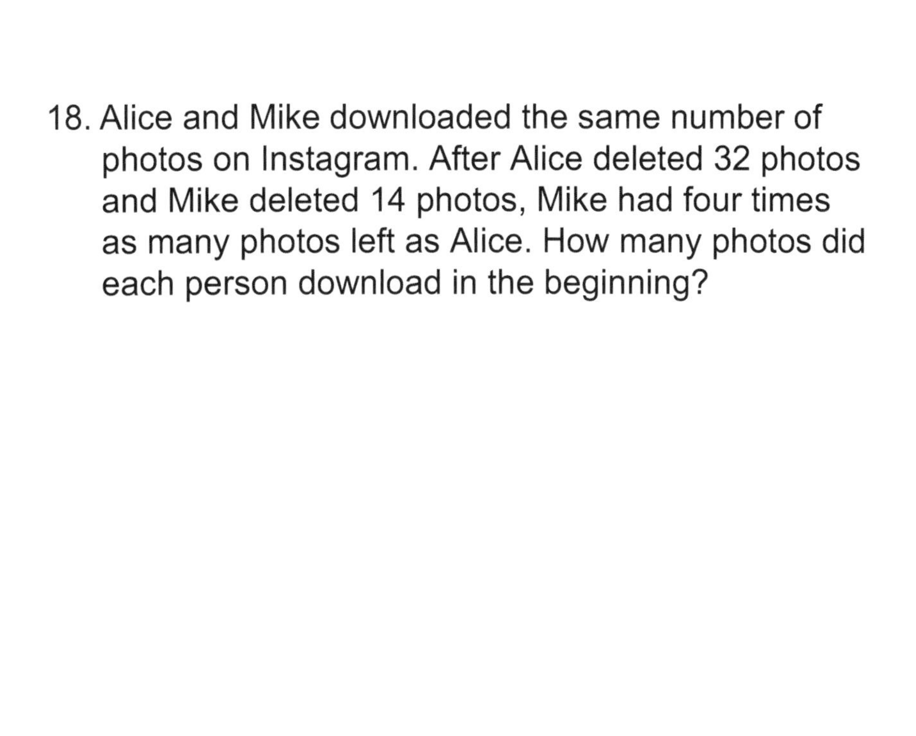

Hint: The difference in the number of deleted photos is equal to the number of times Mike's remaining photos is more than Alice's photos.

19. Ray and Jess had used the same number of tags in their last blog posts to boost traffic. After Ray used another 6 more tags and Jess another 39 more tags, Jess had now used four times as many tags as Ray. How many tags did Ray and Jess use altogether at first?

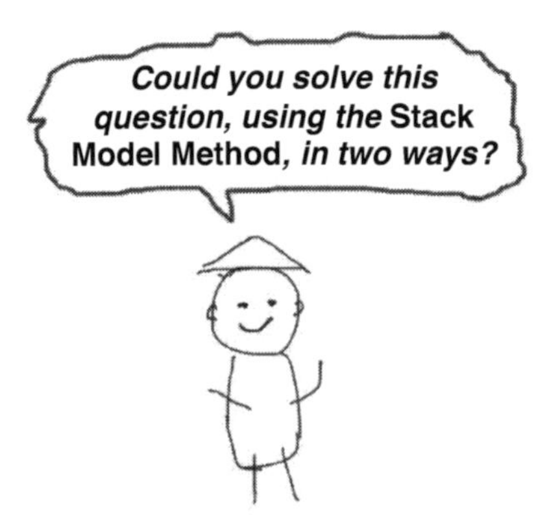

20. Carol and Roy had the same number of Instagram followers. After 6 persons unfollowed Carol and 30 persons unfollowed Roy, Carol had now 5 times as many Instagram followers as Roy. What was the total number of Instagram followers Carol and Roy had at first?

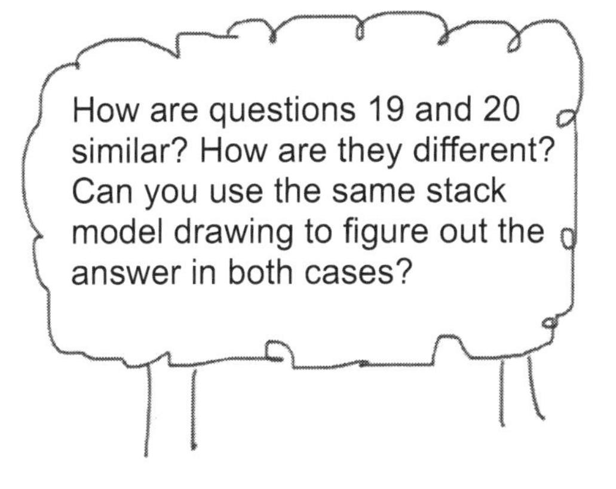

21. Aaron, Beth and Chantal set off on a trip to pick up mangoes. Aaron picked 3 kg more than Beth but 2 kg less than Chantal. If Beth picked 3/4 of the amount that Chantal picked, how much did the three friends pick in total?

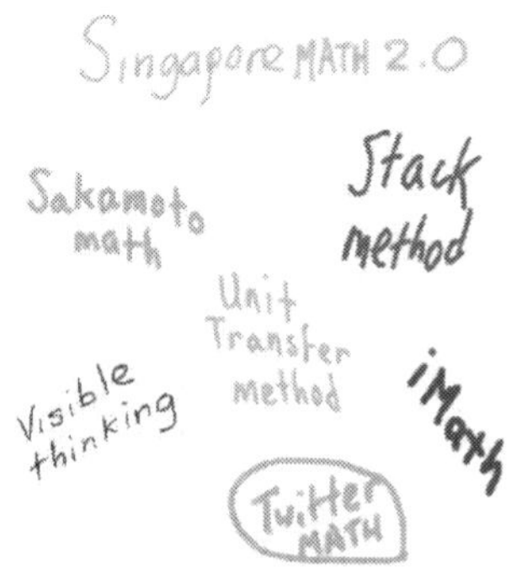

22. A cow weighs 124 kg more than a dog. A goat weighs 96 kg less than the cow. Altogether the three animals weigh 392 kg. What is the mass of the cow?

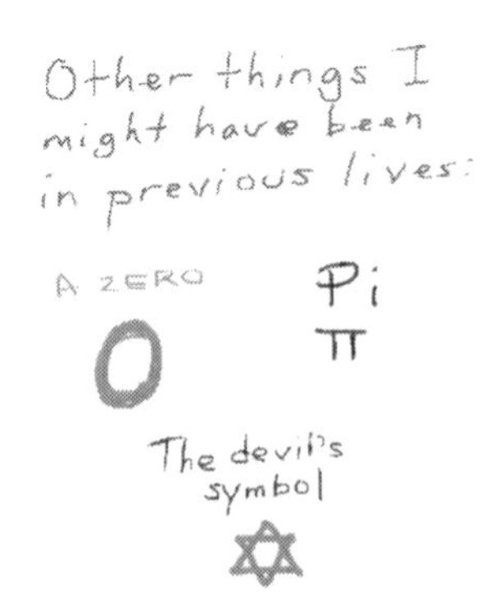

Hint: The basic unit needn't always represent the smallest quantity.

23. There were 3 times as many men as women in a meeting. After 25 men left and 30 women entered, there were 2 times as many women as men. If 69 adults remained, how many adults were in the meeting at first?

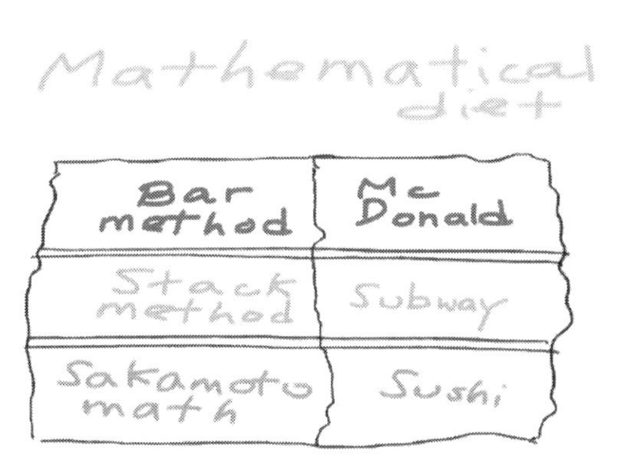

Hint: See Worked Examples 29 and 30.

24. There are a total of 24 mosquitoes and scorpions. The total number of legs is 170. How many mosquitoes and how many scorpions are there? (A scorpion has 8 legs; a mosquito has 6 legs.)

To Stack
OR
Not to Stack

VISUAL LITERACY

Hint: See Worked Examples 12 and 13.

25. Four apps and two e-books cost \$18.
Five e-books and three apps cost \$31.
What is the cost of one app?

iTunes App Store
Christmaths: http://tinyurl.com/7pn8pau

Mathematical Quickies & Trickies
http://tinyurl.com/mxgrn3n

More Mathematical Quickies & Quickies
http://tinyurl.com/py7kcrm

Geometrical Quickies & Trickies
http://tinyurl.com/mjusstm

Hint: See Worked Examples 21 and 23.

Answers & Solutions

Encourage pupils to solve each question in as many ways as possible, besides using the *Stack Model Method*. Tickle the pupils to come up with a different stack model, other than the one(s) given in this book.

Remember: *Not all stack or bar models are created equal — some are more elegant than others.*

1. 120 sheep.

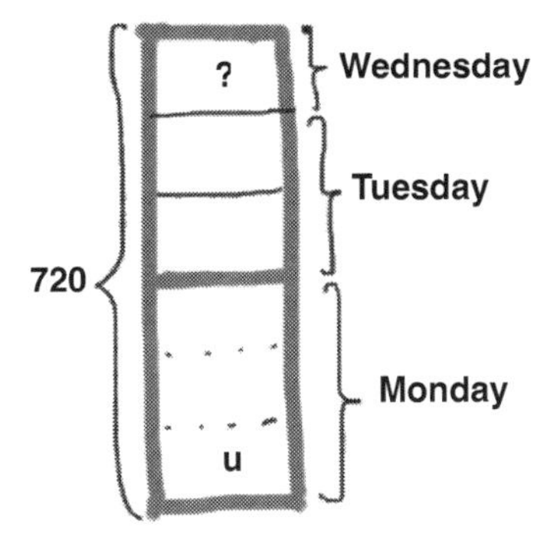

From the model,

6u → 720
u → 1/6 × 720 = 120

120 sheep are left to be sheared on Wednesday.

2. 10 kg.

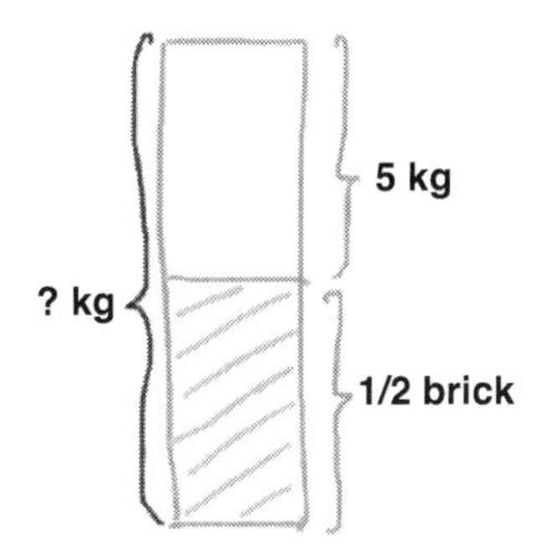

From the stack model,

1/2 brick → 5 kg
1 brick → 2 × 5 = 10 kg

Each brick weighs 10 kg.

3. 15 metres.

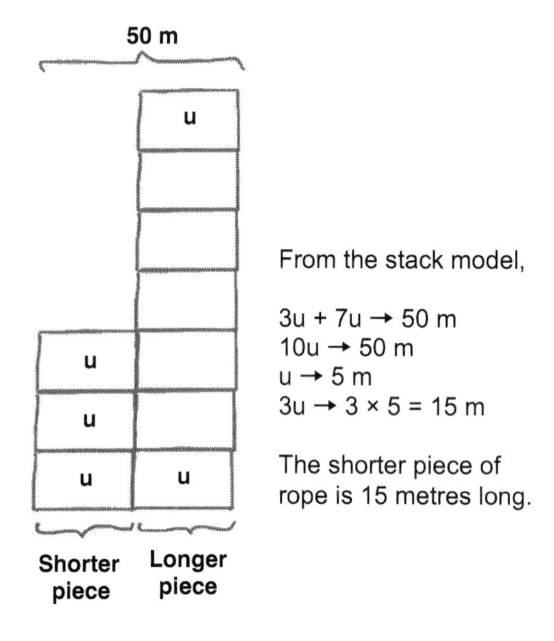

From the stack model,

$3u + 7u \rightarrow 50$ m
$10u \rightarrow 50$ m
$u \rightarrow 5$ m
$3u \rightarrow 3 \times 5 = 15$ m

The shorter piece of rope is 15 metres long.

4. 1/4.

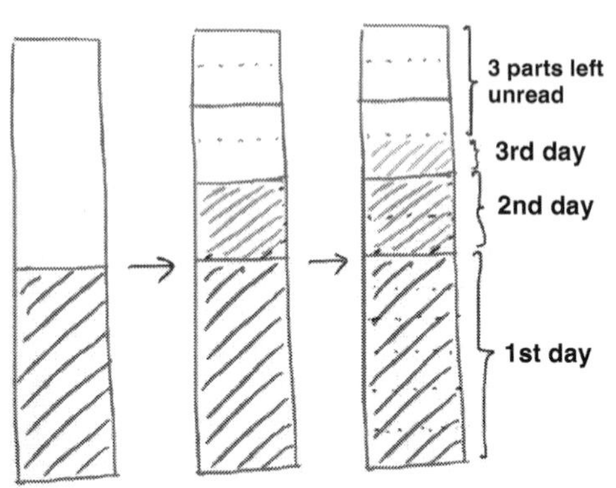

From the stack model,
3 out of 12 parts remained unread.

$3/12 = 1/4$
So, 1/4 of the book remained unread.

5. 2.5 kg.

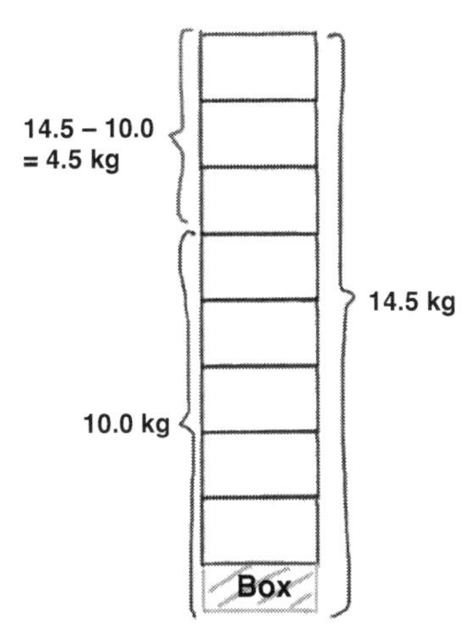

From the stack model,

mass of 3 balls = 14.5 kg – 10.0 kg = 4.5 kg
mass of 1 ball = 4.5 kg ÷ 3 = 1.5 kg
mass of 5 balls = 5 × 1.5 kg = 7.5 kg
mass of the box = 10 kg – 7.5 kg = 2.5 kg

The box weighs 2.5 kg.

6. 35 milk bars.

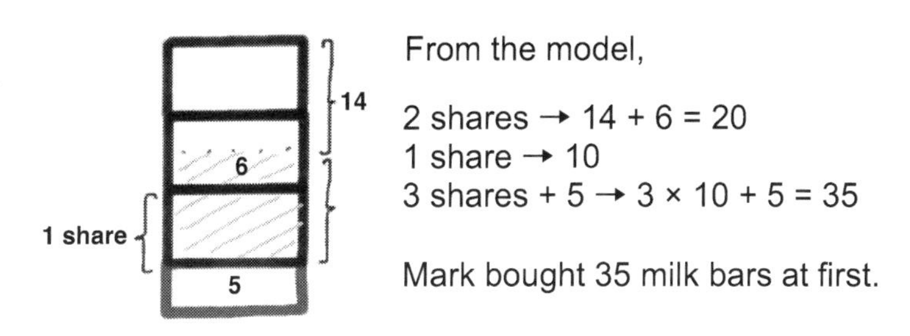

From the model,

2 shares → 14 + 6 = 20
1 share → 10
3 shares + 5 → 3 × 10 + 5 = 35

Mark bought 35 milk bars at first.

7. 2/5.

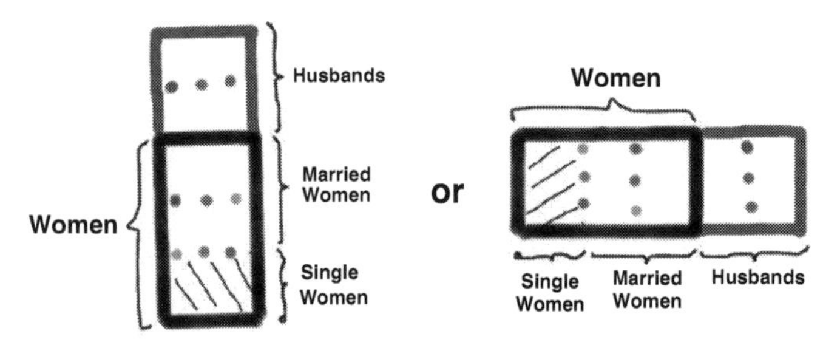

From the model,

if 1 part represents single women, then
2 parts represent married women, and
another 2 parts represent married men.

2 out of 5 parts represent the married men.

2/5 of the people were married men.

8. 720 cm^3.

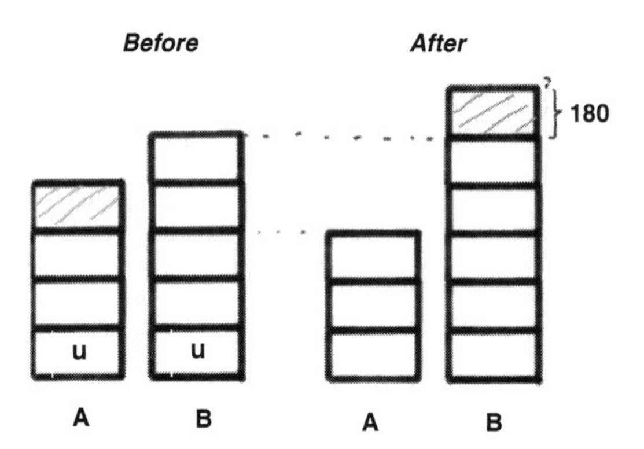

From the model,
u → 180
4u → 4 × 180 = 720

Container A contains 720 cm^3 of water at first.

9. Doris, Esther and Fiona have 55 Facebook fans. They each have more than 10 Facebook fans. Doris has twice as many fans as Esther, and Fiona has the least number of fans. How many Facebook fans does Esther have?

9. 14 Facebook fans.

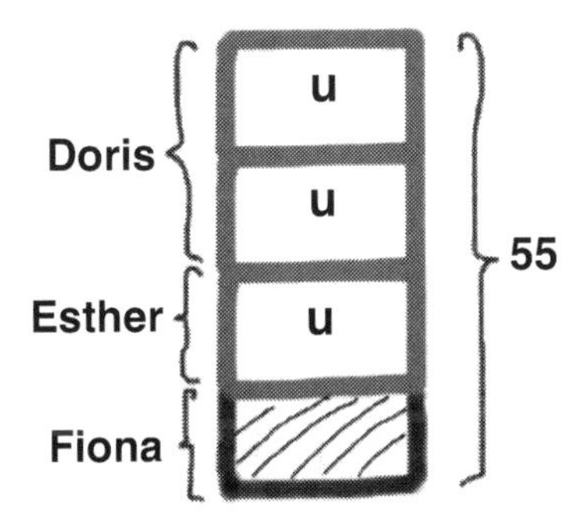

From the model,

if Fiona has 11 fans, then 3u → 55 – 11 = 44—not a multiple of 3;
if Fiona has 12 fans, then 3u → 55 – 12 = 43—not a multiple of 3;
if Fiona has 13 fans, then 3u → 55 – 13 = 42—a multiple of 3.
u → 42 ÷ 3 = 14

So, Esther has 14 Facebook fans.

Or, we may solve the question as follows:

Let F stand for the number of Facebook fans Fiona has.
Then, from the model, $3u + F = 55$ and $F > 10$.

We look for the smallest F that is greater than 10 such that (55 – F) is a multiple of 3. The smallest F is 13.

So, $3u = 55 - 13 = 42$
$u = 42 \div 3 = 14$

10. Leo has 2/3 as many books as Jill. If Jill gives Leo 8 books, they both will have the same number of books. How many books do both persons have altogether?

10. 80 books.

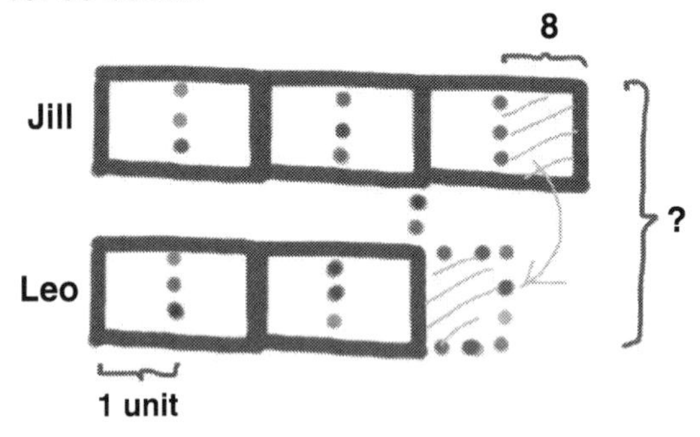

From the model,

1 unit → 8

10 units → 10 × 8 = 80

Both persons have 80 books altogether.

Alternatively, we may use a stack model diagram to work out the answer, as shown on the right.

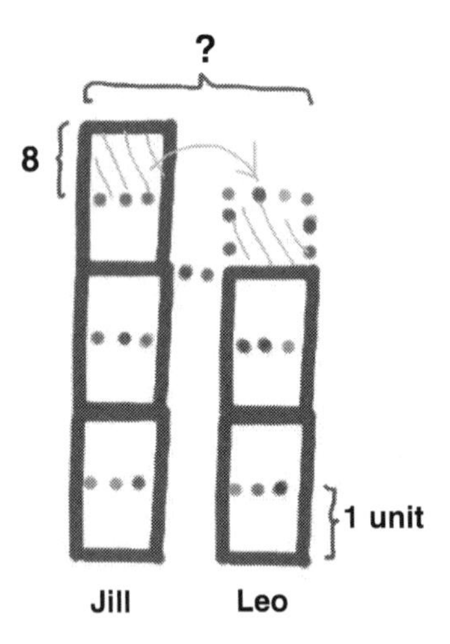

11. 24 people.

Women: 1/3 = 4/12
Girls: 1/4 = 3/12
Men: 1/6 = 2/12

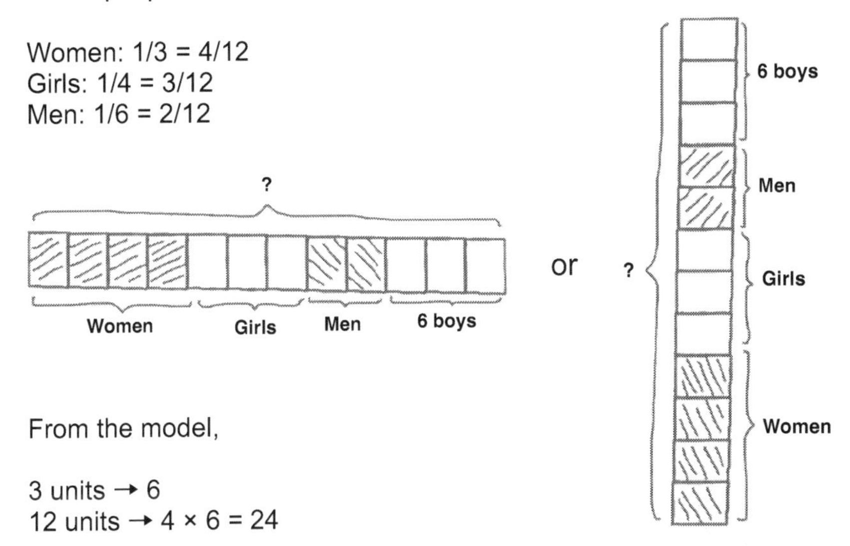

From the model,

3 units → 6
12 units → 4 × 6 = 24

24 people were at the party.

12. $570.

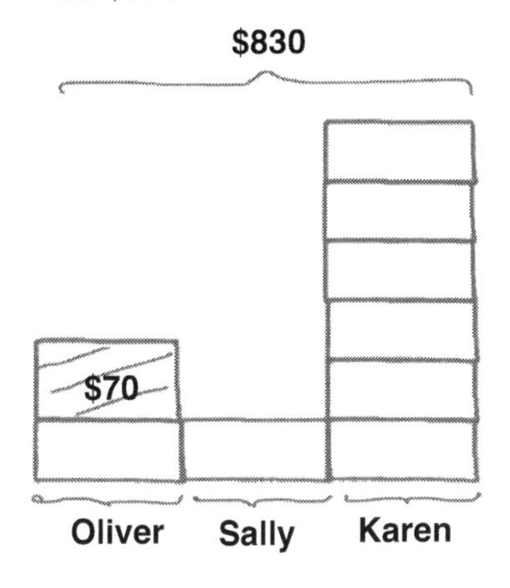

From the model,

8 units + $70 → $830
8 units → $830 – $70 = $760
1 unit → $760 ÷ 8 = $95
6 units → 6 × $95 = $570

Karen saved $570.

13. $36.

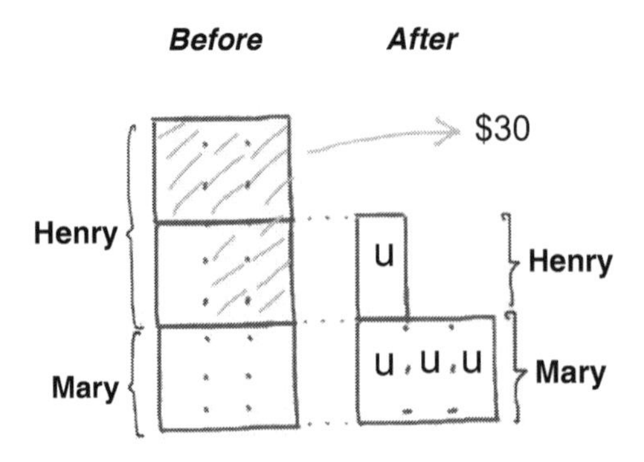

From the model,

5u → $30
u → $30 ÷ 5 = $6
6u → 6 × $6 = $36

Henry had $36 in his PayPal account at first.

14. 14 years old.

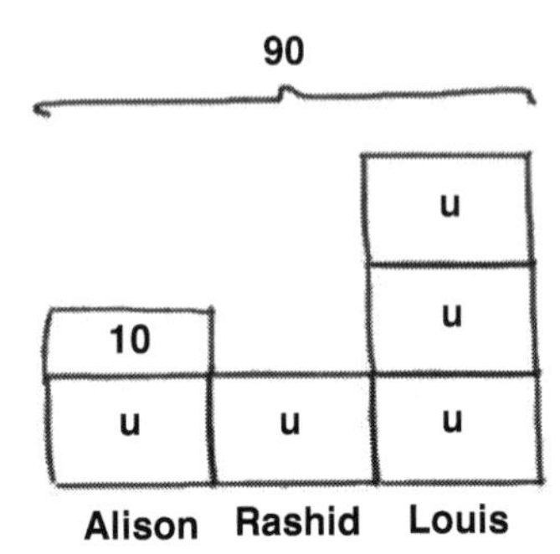

From the model,

5u + 10 → 90
5u → 90 – 10 = 80
u → 80 ÷ 5 = 16
u + 10 → 16 + 10 = 26

Alison is 26 years old now.

26 – 12 = 14
12 years ago, Alison was 14 years old.

15. At present, Joelle is 5 years old, and her mother is 45 years old. How many years from now will Joelle's mother be six times as old as her?

15. 3 years' time.

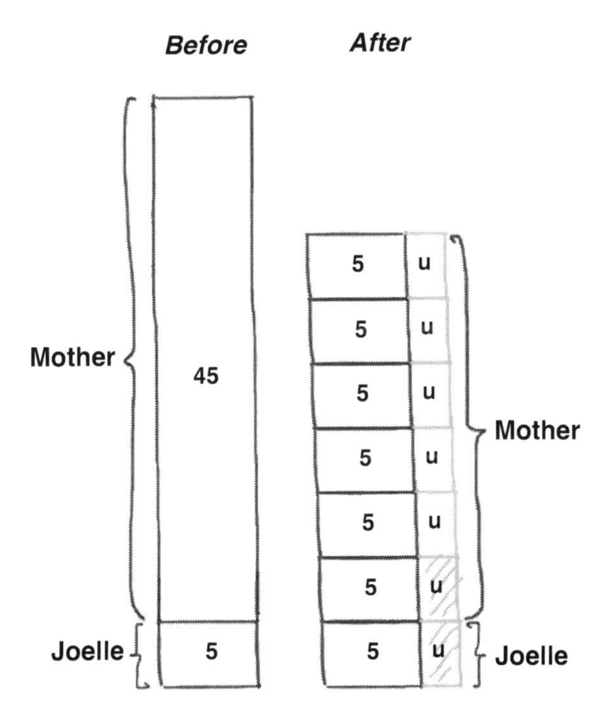

From the model,

$5u + 6 \times 5 \rightarrow 45$
$5u + 30 \rightarrow 45$
$5u \rightarrow 45 - 30 = 15$
$u \rightarrow 15 \div 5 = 3$

In 3 years' time, Joelle's mother will be 6 times as old as her.

16. The sum of the ages of Lisa and Ben is 76. Ben is three times as old as Lisa. In how many years' time will Ben be twice as old as Lisa?

16. 19 years' time.

4 parts → 76
1 part → 76 ÷ 4 = 19
Lisa is 19 years old.
Ben is 3 × 19 = 57 years old.

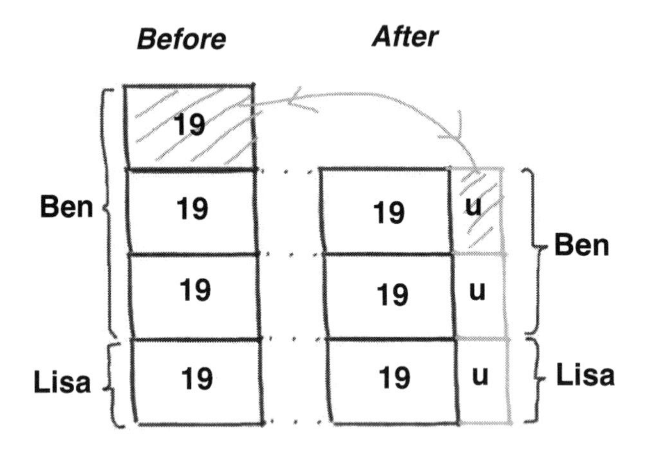

From the model,

u → 19
In 19 years' time, Ben will be twice as old as Lisa.

Check:

	Lisa	Ben
Now	19	57
Later	38	76

76 = 2 × 38

17. 48 years old.

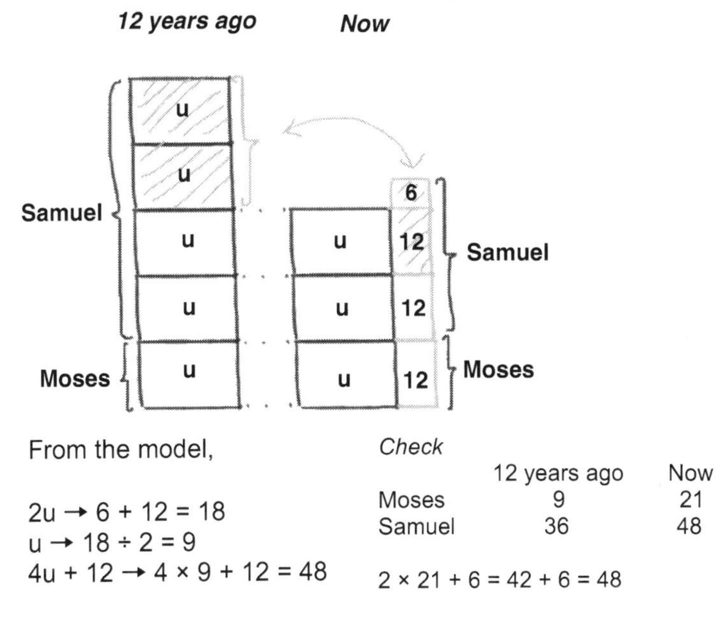

From the model,

2u → 6 + 12 = 18
u → 18 ÷ 2 = 9
4u + 12 → 4 × 9 + 12 = 48

Check

	12 years ago	Now
Moses	9	21
Samuel	36	48

2 × 21 + 6 = 42 + 6 = 48

Samuel is 48 years old now.

18. 38 photos.

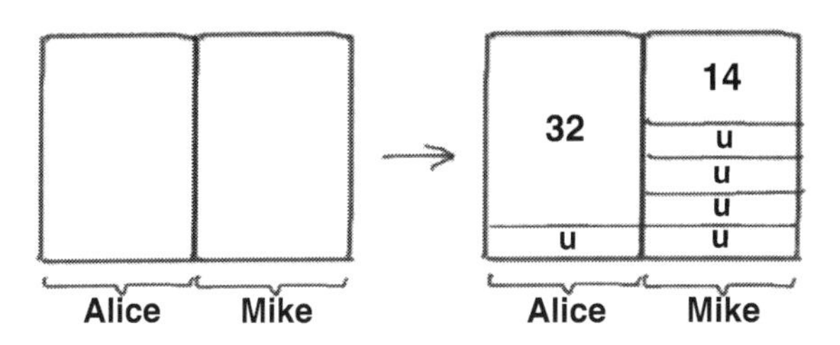

From the model,

3u → 32 – 14 = 18
u → 18 ÷ 3 = 6
u + 32 → 6 + 32 = 38
or
4u + 14 → 4 × 6 + 14 = 38

In the beginning, each person downloaded 38 photos.

19. Ray and Jess had used the same number of tags in their last blog posts to boost traffic. After Ray used another 6 more tags and Jess another 39 more tags, Jess had now used four times as many tags as Ray. How many tags did Ray and Jess use altogether at first?

19. 10 tags.

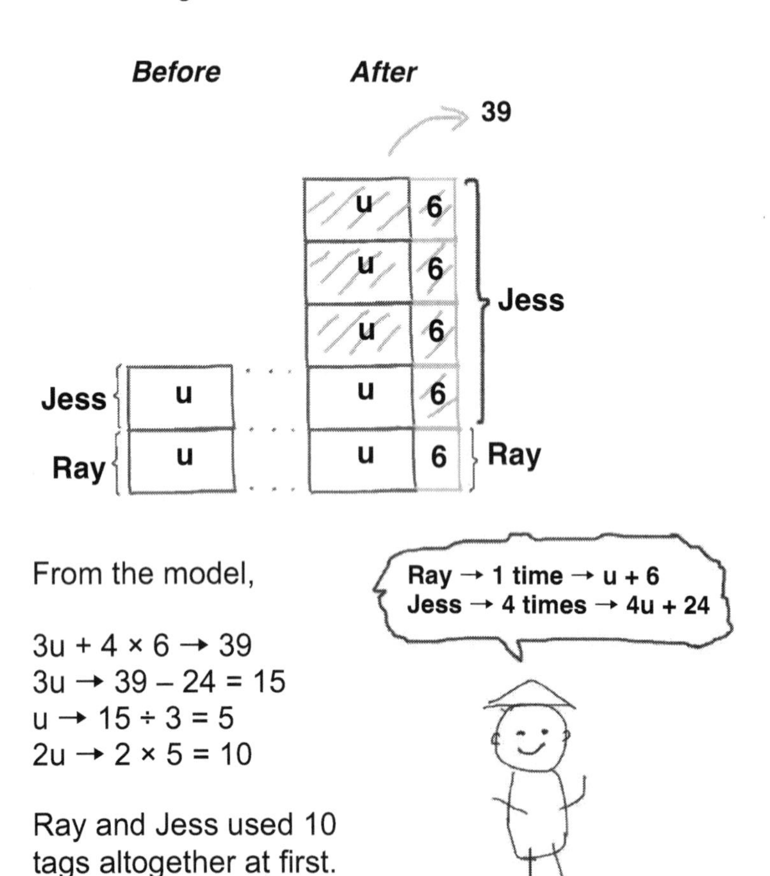

From the model,

3u + 4 × 6 → 39
3u → 39 – 24 = 15
u → 15 ÷ 3 = 5
2u → 2 × 5 = 10

Ray and Jess used 10 tags altogether at first.

20. Carol and Roy had the same number of Instagram followers. After 6 persons unfollowed Carol and 30 persons unfollowed Roy, Carol had now 5 times as many Instagram followers as Roy. What was the total number of Instagram followers Carol and Roy had at first?

20. 72 followers.

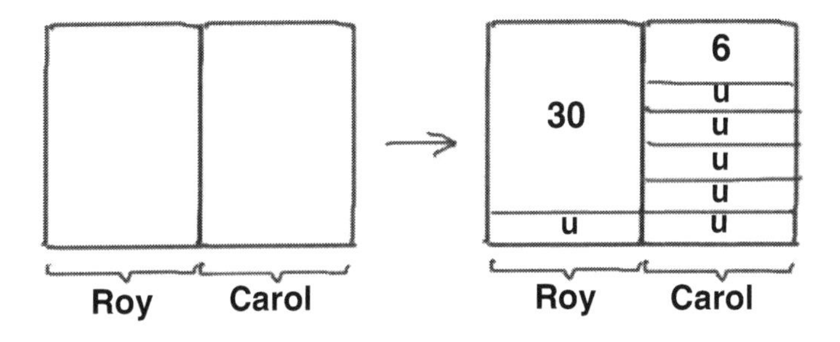

From the model,

4u → 30 – 6 = 24
u → 24 ÷ 4 = 6
u + 30 → 6 + 30 = 36 or
5u + 6 → 5 × 6 + 6 = 36

The total number of Instagram followers Carol and Roy had at first was 72.

21. Aaron, Beth and Chantal set off on a trip to pick up mangoes. Aaron picked 3 kg more than Beth but 2 kg less than Chantal. If Beth picked 3/4 of the amount that Chantal picked, how much did the three friends pick in total?

21. 53 kg.

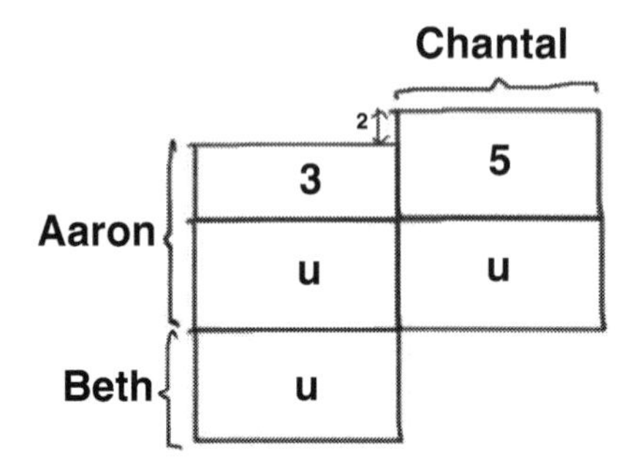

Beth → 3/4 × Chantal
u → 3/4 × (u + 5) = 3u/4 + 15/4
u – 3u/4 → 15/4
u/4 → 15/4
u → 15

Now, u + u + 3 + u + 5 = 3u + 8
3u + 8 → 3 × 15 + 8 = 45 + 8 = 53

The three friends picked up 53 kg of mangoes in total.

22. A cow weighs 124 kg more than a dog. A goat weighs 96 kg less than the cow. Altogether the three animals weigh 392 kg. What is the mass of the cow?

Method 1

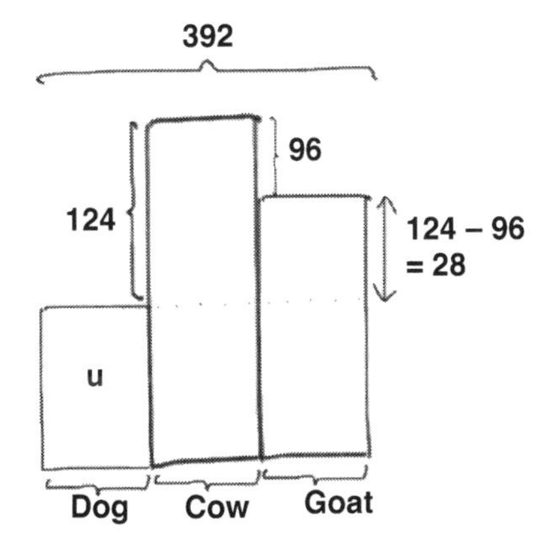

From the model,

3u → 392 – 124 – 28 = 240
u → 240 ÷ 3 = 80
u + 124 → 80 + 124 = 204

The cow weighs 204 kg.

Method 2

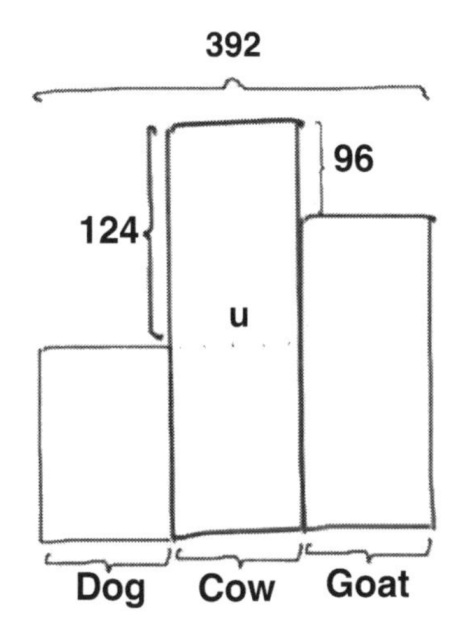

From the model,

3u → 392 + 124 + 96 = 612
u → 612 ÷ 3 = 204

The cow weighs 204 kg.

Note: The basic unit needn't always represent the smallest quantity, as shown in Method 2.

23. There were 3 times as many men as women in a meeting. After 25 men left and 30 women entered, there were 2 times as many women as men. If 69 adults remained, how many adults were in the meeting at first?

Method 1

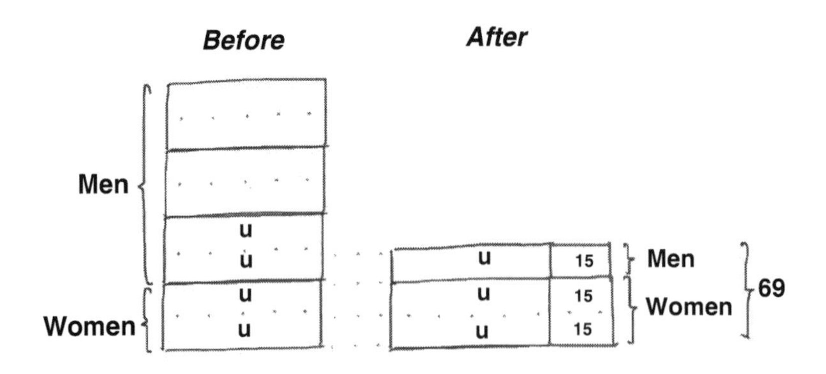

From the model,
$3u + 3 \times 15 \rightarrow 69$
$3u + 45 \rightarrow 69$
$3u \rightarrow 69 - 45 = 24$
$u \rightarrow 24 \div 3 = 8$
$8u \rightarrow 8 \times 8 = 64$

At first, 64 adults were in the meeting.

Check

	Men	Women	Total
Before	6u = 48	2u = 16	64
After	48 – 25 = 23	16 + 30 = 46	69

I wonder why the number 25, which is the number of men who left, isn't used in the solution? Is it redundant?

23. There were 3 times as many men as women in a meeting. After 25 men left and 30 women entered, there were 2 times as many women as men. If 69 adults remained, how many adults were in the meeting at first?

Method 2

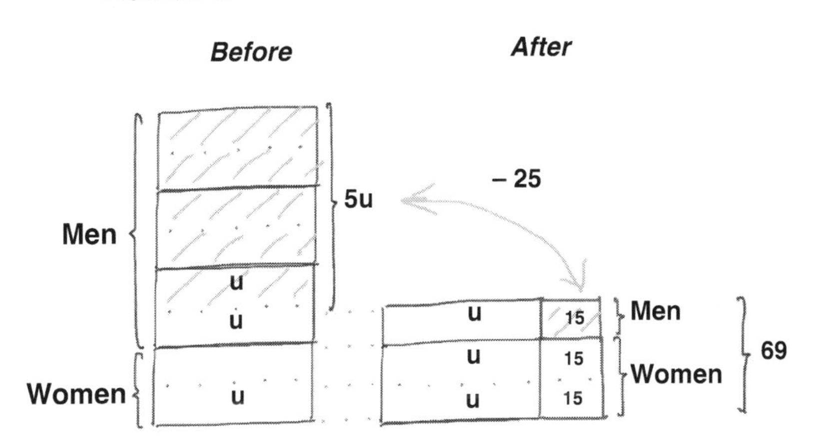

From the model,
$5u \rightarrow 15 + 25 = 40$
$u \rightarrow 40 \div 5 = 8$
$8u \rightarrow 8 \times 8 = 64$

At first, 64 adults were in the meeting.

I wonder why the number 69, which is the number of adults who remained, isn't used in the solution? Is it redundant?

Using Logic

25 men left: – 25
30 women entered: + 30
After 25 men left and 30 women entered, there is a surplus of (30 – 25) = 5 more adults in the room.

Since 69 adults remained, there must be (69 – 5) = 64 adults before.

23. There were 3 times as many men as women in a meeting. After 25 men left and 30 women entered, there were 2 times as many women as men. If 69 adults remained, how many adults were in the meeting at first?

Method 3

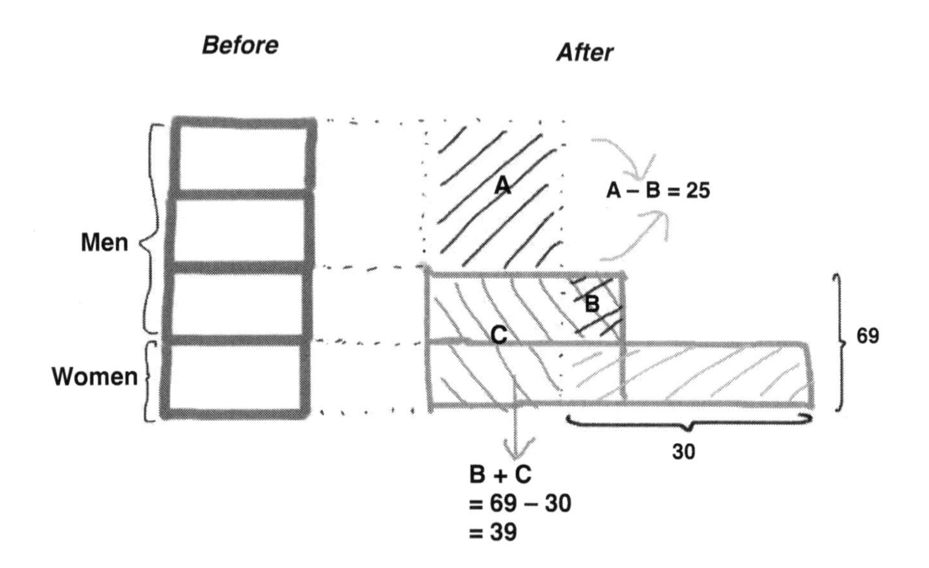

From the model,

number of adults in the meeting at first
= A + C
= (A – B) + (B + C)
= 25 + 39
= 64

Thought Process

Why the following stack model drawing would not be suitable, even though we may theoretically obtain the answer?*

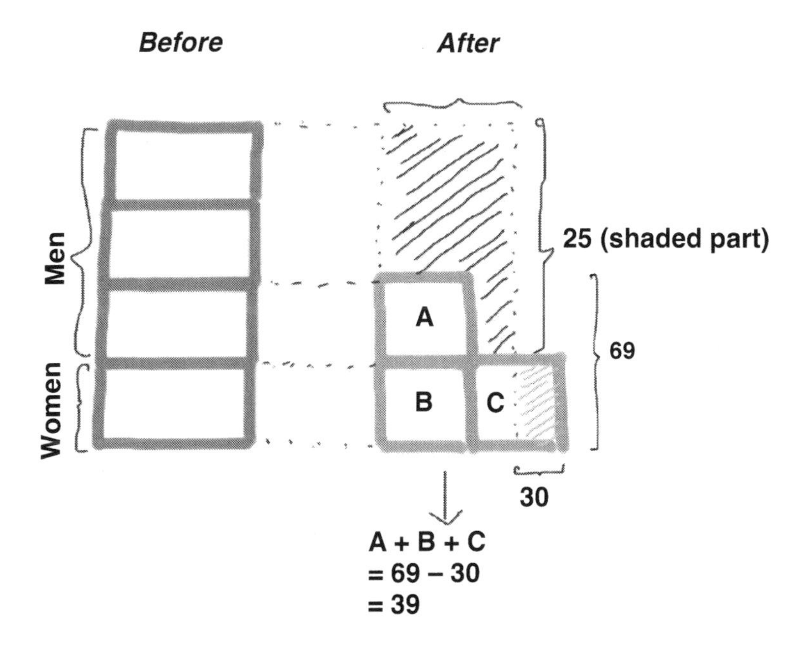

A + B + C
= 69 – 30
= 39

From the model,

number of adults in the meeting at first
= 25 + 39
= 64

* Hint: Are the parts making up the whole reasonably or sensibly proportionate?

24. There are a total of 24 mosquitoes and scorpions. The total number of legs is 170. How many mosquitoes and how many scorpions are there? (A scorpion has 8 legs; a mosquito has 6 legs.)

Method 1

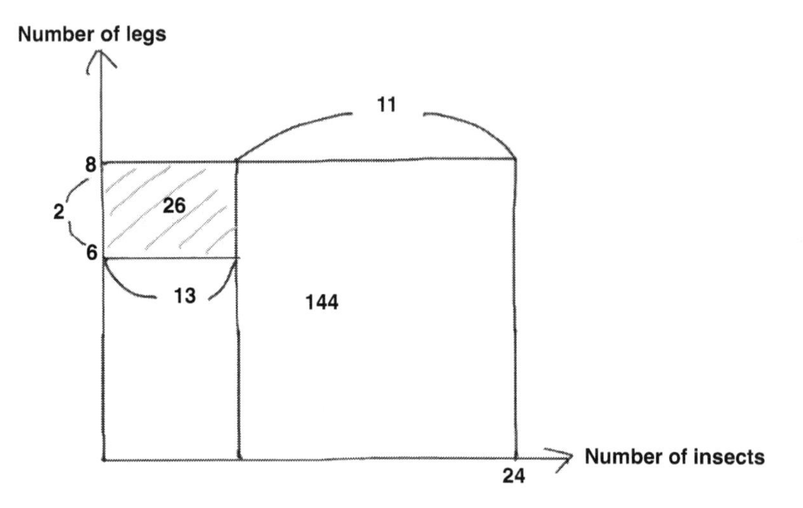

$24 \times 6 = 144$ (if all insects were mosquitoes)

$170 - 144 = 26$
$8 - 6 = 2$ (difference in the number of legs)

$26 \div 2 = 13 \rightarrow$ scorpions
$24 - 13 = 11 \rightarrow$ mosquitoes

There are 13 scorpions and 11 mosquitoes.

24. There are a total of 24 mosquitoes and scorpions. The total number of legs is 170. How many mosquitoes and how many scorpions are there?(A scorpion has 8 legs; a mosquito has 6 legs.)

Method 2

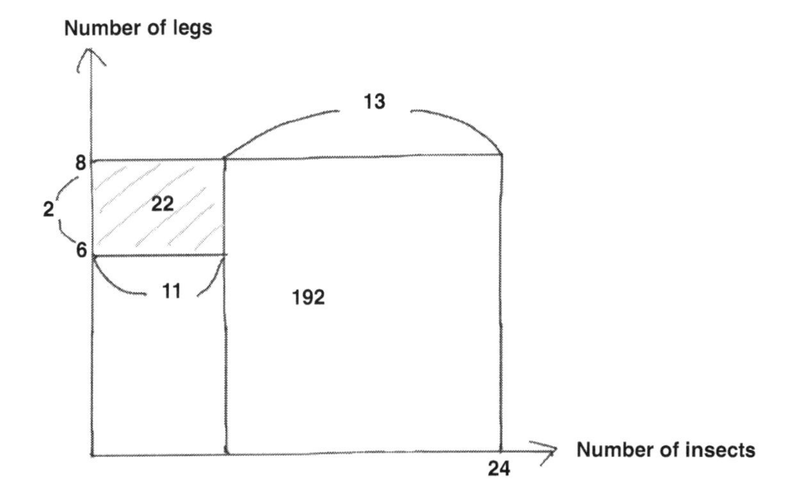

$24 \times 8 = 192$ (if all insects were scorpions)

$192 - 170 = 22$
$8 - 6 = 2$ (difference in the number of legs)

$22 \div 2 = 11$ → mosquitoes
$24 - 11 = 13$ → scorpions

There are 11 mosquitoes and 13 scorpions.

25. Four apps and two e-books cost \$18.
Five e-books and three apps cost \$31.
What is the cost of one app?

4 apps + 2 e-books cost \$18.
3 apps + 5 e-books cost \$31.

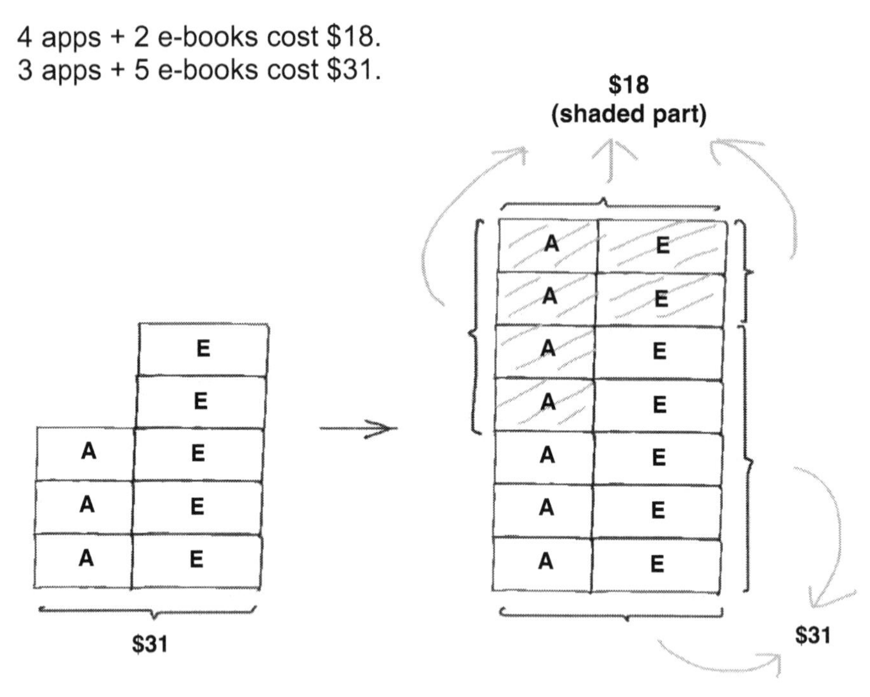

From the model,

7 rows of "1A + 1E" → \$18 + \$31 = \$49
1 row of "1A + 1E" → \$49 ÷ 7 = \$7

1 app + 1 e-book → \$7
2 apps + 2 e-books → 2 × \$7 = \$14
4 apps + 2 e-books → \$18 (given)
2 apps → \$18 – \$14 = \$4
1 app → \$4 ÷ 2 = \$2

One app costs \$2.

Bibliography and References

Tan, S. (2010). *Model approach to problem-solving: Stack & split to solve challenging problems fast!* Singapore: Maths Heuristics (S) Private Limited.

Tan, S. Y. (undated). *Mathswise strategies book upper primary.* Singapore: Horizon Distribution Centre.

Wan, C. H. (2006). *Challenging maths problems made easy.* Singapore: Marshall Cavendish Education.

Yan, K. C. (2014). *Word problems in focus — Primary 4A.* Singapore: Marshall Cavendish Education.

Yan, K. C. (2014). *Word problems in focus — Primary 4B.* Singapore: Marshall Cavendish Education.

Yan, K. C. (2014). *Word problems in focus — Primary 3A.* Singapore: Marshall Cavendish Education.

Yan, K. C. (2014). *Word problems in focus — Primary 3B.* Singapore: Marshall Cavendish Education.

Yan, K. C. (2014). *Primary mathematics challenging word problems (Common Core Edition) — Grade 4.* Singapore: Marshall Cavendish Education.

Yan, K. C. (2014). *Primary mathematics challenging word problems (Common Core Edition) — Grade 3.* Singapore: Marshall Cavendish Education.

About the Author

Kow-Cheong Yan is the author of Singapore's best-selling *Mathematical Quickies & Trickies* series and the co-author of the MOE-approved *Additional Maths 360*. Besides coaching mathletes and conducting recreational maths courses for students, teachers and parents, he edits, ghostwrites and consults for *MathPlus Consultancy.*

Kow-Cheong writes about the good, the bad and the not-so-ugly of Singapore's maths education and of the local educational publishing industry. Read his two maths blogs at **www.singaporemathplus.com** and **www.singaporemathplus.net**, or follow him on Twitter @MathPlus, @Zero_Math and @SakamotoMath.

His e-mail coordinates are **kcyan@singaporemathplus.com** and **kcyan.mathplus@gmail.com**.

Visit Kow Cheong's Facebook Pages and Pinterest Boards at these virtual addresses:

fb.com/SingaporeMathPlus
fb.com/AddMaths360
fb.com/Christmaths
pinterest.com/MathPlus

Printed in Great Britain
by Amazon